THE IDEA OF CONSCIOUSNESS

THE IDEA OF CONSCIOUSNESS

Synapses and the Mind

Max R. Bennett
Institute for Biomedical Research
University of Sydney
Australia

Illustrations by
Gillian Bennett

harwood academic publishers
Australia • Austria • China • France • Germany • India • Japan
Luxembourg • Malaysia • Netherlands • Russia • Singapore
Switzerland • Thailand • United Kingdom

Copyright © 1997 OPA (Overseas Publishers Association) Amsterdam B. V.
Published in The Netherlands by Harwood Academic Publishers.

Amsteldijk 166
1st Floor
1079 LH Amsterdam
The Netherlands

British Library Cataloguing in Publication Data

A catalogue record for this book is available from the British Library.

ISBN 90-5702-203-6

Cover design by Kerry Klinner.

CONTENTS

INTRODUCTION

The present work is an attempt to answer the question: to what extent has neuroscience explained consciousness? It is therefore concerned with describing the events that occur in the brain during conscious experience. These events involve the activation of particular neurons as a consequence of the input of synaptic connections from other neurons. The phenomena that arise due to the interplay of these synaptic connections is then of central interest in this book. It is how the mind, or at least the contents of the mind, arise from the workings of synaptic connections that is to be explored, hence the subtitle *Synapses and the Mind*. Various experimental approaches to the problem are described, and summarized in Chapter 1, so as to give the reader an indication of the range of research on this subject.

The problem of consciousness is a major issue of contemporary philosophy and science. One view, that adopted by the philosopher Daniel Dennett, suggests that we now have *Consciousness Explained* by assuming an equivalence between mental and brain states, an idea that has been popular since it was suggested by Herbert Feigl in the 1950s. On the other hand, the philosopher Thomas Nagel comments that:

> people are now told at an early age that all matter is really energy. But despite the fact that they know what 'is' means, most of them never form a conception of what makes this claim true, because they lack the theoretical background.

> At the present time the status of physicalism is similar to that which the hypothesis that matter is energy would have had if uttered by a pre-Socratic philosopher. We do not have the beginnings of a conception of how it might be true. In order to understand the hypothesis that a mental event is a physical event, we require more than an understanding of the

word 'is'. The idea of how a mental and a physical term might refer to the same thing is lacking, and the usual analogies with theoretical identification in other fields fail to supply it.

This identity problem is particularly acute when considering qualia; that is, the feelings and sensations that accompany our conscious awareness of the world. The philosopher John Searle is well known for his view that qualia cannot be treated in this way, a proposition that is examined from a neuroscience perspective in Chapter 2.

Several biologists have recently turned their attention to the problem of how the workings of synaptic neural networks in the brain could give rise to the contents of conscious mental states, with the assumption that these are equivalent as has been claimed by Francis Crick in *The Astonishing Hypothesis*. Gerald Edelman assumes much the same position in *Theory of Neuronal Group Selection or Neural Darwinism*. Without necessarily adopting such a viewpoint, Chapters 3 and 4 consider what states of synaptic networks might accompany awareness of phenomena in consciousness.

It is at least clear that the synaptic connections in the neural networks of the brain are pertinent to both consciousness and to the contents of that consciousness. This point is emphasized in Chapter 5 by examining the workings of the brain that accompany those distortions of consciousness which occur in a mental illness such as schizophrenia. The consideration of such unusual conditions of mental life leads naturally to the question of whether it is possible to ascertain objectively if animals other than ourselves have consciousness, and if so what form this might take in different species, a problem examined in Chapter 6.

The final chapter returns to the question of how the workings of the brain give rise to such extraordinary phenomenon of consciousness as qualia, even given the action of synaptic connections that have been elucidated in the previous chapters. The neuroscientist John Eccles and more recently the mathematician Roger Penrose have argued that the synaptic connections in the brain must function in such a way as to involve quantum mechanical principles. Eccles sees this as a way of bridging between a physical world of synaptic networks and a world of the mind, as is explained in Chapter 7, whereas Penrose regards this in a large theoretical framework involving solution to the problem of how to give a coherent account of the meaning of quantum mechanical principles. Very recently, the philosopher David Chalmers has emphasized in *The Conscious Mind* that it does not matter if quantum mechanical principles are involved in synaptic function or not, as the physical world as presently understood is a closed system of explanations that is therefore sealed off from the world of consciousness. The present work is then doomed to failure in respect to offering insights into the problem of consciousness because it takes a neuroscientist's perspective. However, the history of science and philosophy shows that the experimental probing of natural phenomena frequently reveals unexpected surprises. These may profoundly change our way of thinking about subjects that seemed opaque to further understanding, such as was once the case with the nature of matter and that of the living state. Only time and ingenuity will tell if consciousness can be explained by analysis of the synaptic connections in the brain. In the meantime, neuroscientists will have great pleasure in solving important puzzles, with some of these having profound implications for the alleviation of mental suffering and how we view our species.

Most of this work was first published as a series of occasional essays intended to satisfy the curiosity of those interested in the origins of consciousness. I hope the additional material helps to more closely fulfil that aim. To this end, the books mentioned above will be found useful. To assist the reader further there are references to general works of both neuroscience and philosophy that provide background for the contents of each of the chapters. More specialized references to research papers are also listed for those who might wish to enter into the exciting experimental world that is now opening up through the study of synaptic connections in the brain.

CHAPTER 1   AN INTRODUCTION TO
CONSCIOUSNESS AND THE BRAIN

1.1 Erwin Schrodinger's *What is Life?*

Working in Dublin in 1944, during the Second World War, Erwin Schrodinger published a book which had an enormous influence on the history of biological science in the second half of the twentieth century (Figure 1.1). In 1926 Schrodinger, together with Heisenberg, had created the theory of quantum mechanics[1]. Towards the end of the 1930s Schrodinger turned his mind to the question of the physical origins of humans. He started to consider the structure of genes and the extent to which human development occurs as a consequence of the unfolding of the information in genes. What the genetic material was made of was not known at the time that Schrodinger wrote his book[2] *What is Life?*. The guess at that stage was that it was made of protein and it was only shortly after the appearance of Schrodinger's book that Oswald Avery, working in the Rockefeller Institute in New York, discovered that the genetic material was made of deoxyribonucleic acid (DNA). Schrodinger's little book bridged the revolution in physics that occurred in the first part of the century with the new revolution that was to dominate the second part of the century, namely molecular biology. In his book, Schrodinger conjectured that there must be a code in the genes which is read out by the cell as a set of instructions that guide its differentiation so as to give rise to the structure of the human embryo and therefore to further development. In this way, Schrodinger developed the concept of a genetic code, one of the most fruitful ideas in biology. Furthermore, he speculated that the structure of the genetic code should be susceptible to physical analysis by means of such new techniques as crystallography. Crystallography had been developed as a science in the United Kingdom by the Braggs, father and son; they were from Adelaide and had won the Noble Prize together in 1915.

1

Figure 1.1   Erwin Schrodinger (1887–1961).

Figure 1.2   Francis Crick (born in 1916).

After the Second World War, some of the brightest physicists decided not to go into theoretical physics but instead to try their hand at biology. The end of the Heroic Period in physics, which had been dominated by quantum mechanics, occurred in 1948 when Richard Feynman, working at Cornell, developed the theory which is referred to as quantum electrodynamics; this provides a description of how light interacts with matter. In Chapter 7 the possibility that quantum mechanical effects are involved in the origins of consciousness will be considered. Only a year after Richard Feynman had developed his theory of quantum electrodynamics, a physicist named Francis Crick began working with the Braggs in Cambridge (Figure 1.2). Schrodinger's book had a major influence on Crick making this move. In 1951, together with a young American James Watson, Crick applied the concepts of crystallography to the task of unravelling the detailed structure of DNA itself. In this way, he furthered the research program laid down by Schrodinger some five years earlier. To some extent the famous paper by Watson and Crick, published in *Nature* in 1953, completed the research aims of Schrodinger's book by indicating that the structure of the DNA molecule held within it the clues to its own replication. The last forty years of biological science have been largely dominated by molecular biology, which has as part of its foundations these observations by Watson and Crick[3].

Crick's work, like that of Schrodinger before him, is likely to have far reaching effects on the future of research as young scientists see it[4,5]. In *The Astonishing Hypothesis*[5] Crick suggests that the great challenge for science in the next century is not to be found in quantum mechanics, nor in molecular biology, but in the understanding of what it is that develops in the brain of a human embryo that gives rise to consciousness. Figure 1.3 shows a human embryo at five weeks after conception. The hands and legs are already formed but there is only a hint of the digits. The body is clearly connected to an umbilical cord. The brain is developing above the eyes. The spinal cord is also clearly delineated. The adult brain has some 100 thousand million neurons, many of which have about 10 thousand connections (or synapses) on them from other neurons. This results in some 100 million million synapses in the brain. As these synapses are capable of modifying their properties according to experience, there is an extraordinary range of possibilities in the synaptic wiring of the brain.

1.2 The problem of consciousness approached through vision

What is this phenomenon of consciousness? In order to tackle consciousness it is necessary to look at the human brain, at its structure and function. Perhaps the best path to follow in the quest to understand consciousness is offered by the problem of visual perception. This is because most humans, and other primates,

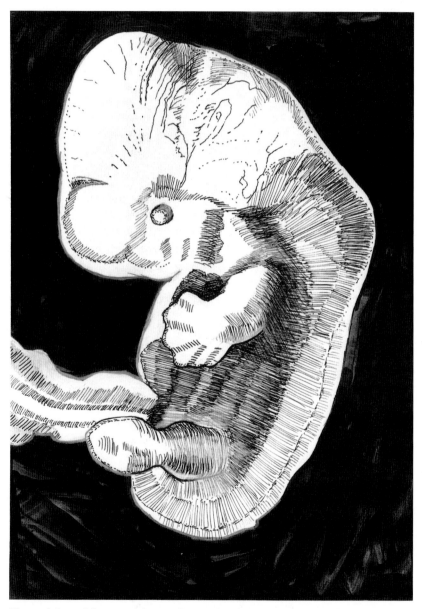

Figure 1.3   A human embryo at five weeks after conception (11 mm long).

receive much of their sensory information via the visual system and so consciousness is often closely linked with visual perception.

Figure 1.4 delineates the principal pathways in the higher levels of neocortical
function concerned with the identification of objects and their movement in space.
A coloured triangular form is observed moving in the visual field. The information about color and form of an object on the one hand and about the movement
of the object on the other are coded for separately within the retina of the eye, to
be sent in parallel pathways along separate sets of axons for further processing.

The identity of objects and their color is taken up by a set of retinal ganglion
cells and conveyed via the structure called the dorsal lateral geniculate of the
thalamus, which acts as an interface between the external world and the neocortex,
to the primary visual (or striate) cortex in the occipital lobe at the back of the
head. From there it is projected down into the temporal lobe and it is here that the
identification of something as a face is carried out. In the medial temporal area
(MT) just in front of the striate cortex, the coded information undergoes a trans-
formation into the elements that recognizably belong to the object. Identification
of an object is completed in the inferior temporal cortex, where the reconstruction of the complete object and its color is carried out. Here specific neurons fire
when the object is viewed (as explained in more detail in the discussion relating
to Figure 1.5).

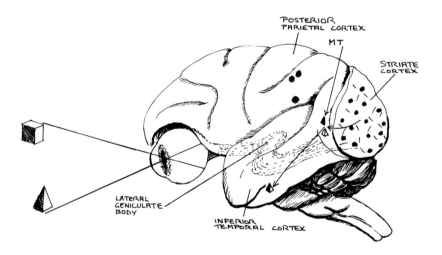

Figure 1.4   The main pathways in the brain concerned with the visual
identification of objects and their location and movement in space.

The other main projecting pathway is concerned with determining where an object is located in visual space. To a large extent, this is dependent on information regarding the motion of objects in space and so the other parallel line of coded information is conveyed from the retina through the thalamus up to the visual cortex within the occipital lobe and into the posterior parietal cortex, also in front of the striate cortex, where particular neurons fire in relation to the movement of the object.

These two pathways, going either to the temporal lobe or to the parietal cortex, result in the holistic experience of seeing an object which moves.

To what level of sophistication can the temporal lobe identify an object? In order to answer this, electrodes have been placed into the temporal lobes of monkeys to determine whether there are neurons which fire maximally when a monkey is viewing a specific object. In a monkey's temporal lobe there are neurons which respond specifically to the presentation of faces; indeed, these neurons discharge specifically depending on whether the faces are presented in profile or face on. A series of images and the neuron firing rates (given by the black vertical bars) which result when the monkey looks at these images are depicted in Figure 1.5. When the monkey is looking directly at the image of another monkey, face on, we get maximal firing but when the image of the head gradually turns around so that it only appears in profile, then the firing occurs at a much lower rate. And although somewhat 'monkey like', if the picture of a

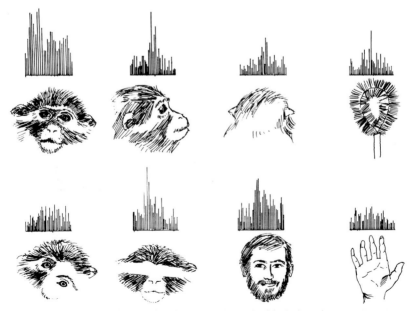

Figure 1.5   Firing rate responses to specific images.

toilet brush is presented, then the rate of firing of the neurons is much less than when the image of another monkey face on is presented. Also, if an image is presented consisting of the juxtaposition of different elements of the face in a bizarre geometry, again the firing rate is not nearly as high as it is when those elements are put together to make up a proper monkey face. Furthermore, it can be shown that if different elements of the monkey's face such as the mouth or the eyes are taken away, then the firing rate will drop. Finally, the pattern corresponding to presentation of a monkey's face is different to that resulting from presentation of a human face. So there are very specific neurons in this temporal lobe which are specific in the sense that they will fire vigorously only when a particular kind of object, in this case a particular face, is presented.

1.3 The origin of holistic experiences in consciousness

The next consideration is that neurons might exist in the temporal lobe for identifying faces, but the experience of whether a face is moving across a visual field is not generated in the temporal lobe but in the parietal cortex. As indicated in Figure 1.4, the question now is how an holistic experience is generated in consciousness when the face is moving across the visual field.

The main difficulty in experiencing a particular object from amongst the many different complex impressions that impinge on the retinas involves the linking together of related features that constitute the object. This is referred to as the binding problem, in which the constituent visual elements of the object must be bound together in the visual cortex so that identification of the object as separate from the other components in the visual image is made.

A very important discovery concerning this problem was recently made by two groups in Germany: Wolf Singer and his colleagues as well as R. Eckhorn and his colleagues. They found that neurons firing in one part of the brain, concerned for example with the movement of an object such as in the parietal cortex, and neurons firing in a different part of the brain, such as in the temporal cortex concerned with the fact that the thing that is moving is a face, are both firing at a rate modulated at a frequency of about 40 to 50 Hz. Furthermore, they fire in phase together and not asynchronously. This 'phase-locked' firing of the different assemblies of neurons which are activated by the same visual object (even though they are located in different parts of the cortex) gives a 'binding together' of the different visually related aspects of the scene. It is only those neurons which are firing in phase with this 40 to 50 Hz frequency that are giving an experience of attending to a particular part of our visual field. The neurons subserving those parts of the image on the cortex that are not being attended to are not firing with a modulation of 40 to 50 Hz and are not firing in phase. This problem of how an holistic visual experience is generated is considered in some detail in Chapter 3.

To some extent it can be claimed that this is all very interesting but that the problem has merely been pushed back one step. Consciousness of phenomena might be associated with sets of particular neurons in the brain firing in phase at 40 to 50 Hz, but how is it that neurons firing off in this pattern give rise to experiencing or attending to only a particular facet of the visual world? After all, it is only a particular facet of the visual world which is entering consciousness. What is there particular about the neurons firing at 40 to 50 Hz and in phase which allows the experience at the conscious level of those events which are coded for by these particular neurons? Perhaps this is as much as neurophysiology can offer in understanding the phenomenon of consciousness. All that science will illuminate in relationship to consciousness of the world may be that sets of neurons subserving certain specific kinds of functions fire in a certain way. All that sets of neural networks can do, according to this view, is solve algorithmic problems, a subject that is taken up in Chapter 2. The question of what it is that links consciousness to these sets of neurons may not be one which science can answer.

1.4 The loss of specific elements of consciousness through disease and stroke

If there is any interference at all with normal brain function, for example through disease or injury to the inferior temporal lobe or through inappropriate development, then the capability for consciousness in those areas of experience normally subserved by the injured neurons is lost. A deeper knowledge of the nervous system will lead to the alleviation of neurological disease, whether this occurs through developmental malfunction, through injury or because of a particular virus as probably occurs in multiple sclerosis.

Hypertension leads to a breakdown of the vasculature and hence to vascular stroke resulting in the destruction of certain parts of the brain. One area in particular in which hypertension most commonly leads to stroke is in the temporal lobe where information is gathered about the identification and consciousness of objects. Figure 1.6 shows a self-portrait by an artist, Anton Raderscheidt, who suffered a vascular stroke affecting the right hemisphere of the brain, in particular the parietal cortex. Shown here are four self-portraits that he painted at different stages of recovery. The parietal cortex is not only concerned with location and movement but also subserves the process of attention. There is a mechanism in the parietal cortex that determines which sets of neurons will fire off in phase at 40 to 50 Hz throughout the rest of the cortex. The stroke had the effect of blocking the attentional mechanisms in the inferior temporal lobe on one side of the brain so that the painter was not able to recognize one side of his face when he looked in the mirror. This did not occur because there had been direct injury to the temporal lobe. It most likely occurred because the attentional mechanism in the parietal cortex on one side of the brain could not determine that the temporal lobe on that side should contain neurons that fired in phase at 40 to 50 Hz. As a consequence, when he was asked to paint a portrait of himself the painter ignored that side of his body which was no longer attended to because of the stroke. As his condition improved, successive self-portraits gradually reconstituted the entire image until finally, after a year, he was able to attend to the entire aspect of his face, although recovery was not complete and he died several years later.

The fact should be emphasized that the reason the painter was unable to 'see', as it were, one side of his face and body when he did a self-portrait a few weeks after the stroke was not due to any injury to the visual pathway. What had most likely been injured was the mechanism in the parietal cortex which determines the setting up of 40 to 50 Hz in phase firing of neurons which then allows consciousness to be expressed. An easy experiment could be carried out to show that he could see the other side of his body. All that had to be done was to block off that side of the painter's visual field which allowed him to see the side of his body which he could normally paint. He would then paint the otherwise ignored side of his body and face. Some people who have a lesion of this kind become emotionally upset when they are forced to attend to that part of their body that they don't normally recognize as being there. They regard that part of their body as foreign. It is as if they have a Siamese twin attached to them which they do not want to know about.

Injuries to the temporal cortex give rise to epileptic seizures which may be accompanied by different kinds of psychological disturbances. As shown in Figure 1.7, the common form of epileptic fit occurs as a focal seizure originating in a localized region of the temporal lobe, often referred to as the limbic system. These seizures involving the temporal lobe are accompanied by visual, auditory and olfactory hallucinations. They also involve the vivid recall of memories, because the limbic system includes the hippocampus which is required for the laying down of new memories. For example, injury to the temporal cortex may lead to hallucinations that involve seeing people or objects that aren't actually present. This is due to epileptic discharges in the neurons of the temporal cortex which are normally activated by a particular image but start to fire despite the fact that the image is not present.

Other forms of temporal lobe epileptic activity occurring as a consequence of stroke in the temporal lobe area are seizures which give rise to the perception of sudden enlargement and distortion of the object being viewed. So hallucinations are not limited to perceptions of events occurring which are not occurring at all, but include distorted perceptions of real events.

Figure 1.6   Defects in the ability to attend to parts of one's own body following a stroke.

Figure 1.7 Localization of focal epileptic seizures in the brain.

Clinical type		Localization
1	Somatic motor:	
	Jacksonian (local motor)	Prerolandic gyrus
	Masticatory	Amygdaloid nuclei
	Simple contraversive	Frontal
2	Somatic and special sensory (auras):	
	Somatosensory	Postrolandic
	Visual	Occipital or temporal
	Auditory	Temporal
	Vertiginous	Temporal
	Olfactory	Mesial temporal
	Gustatory	Insula
3	Visceral:	Insuloorbital-frontal
	Autonomic	cortex
4	Complex partial seizures:	
	Formed hallucinations	Temporal
	Illusions	Temporal
	Dyscognitive experiences (déju vu, dreamy states, depersonalization)	Temporal
	Affective states (fear, depression, or elation)	Temporal
	Automatism (ictal and postictal)	Temporal and frontal
5	Absence	'Reticulocortical'
	Bilateral epileptic myoclonus	'Reticulocortical'

Source: Adapted from Penfield and Jasper (1954) *Epilepsy and Functional Anatomy of the Human Brain.* Boston: Little Brown.

Thus the most common human epileptic condition is not that of generalized seizures, which start out in the entire neocortex of grey matter as in 'petit mal' or 'grand mal' epilepsy, but of focal seizures in the limbic system.

Hallucinations are also associated with schizophrenia. In this case, subjective phenomena apparently occur that are more real than sets of phenomena which are actually occurring, a subject examined in detail in Chapter 5.

1.5 Neuronal growth factors for the regeneration of nerve connections

To what extent will we be able to bring alleviation of the symptoms of stroke and of diseases which afflict the inferior temporal lobe? This requires focusing on a number of different technologies which have been initiated in the last few years. They will be dealt with in some detail here, instead of devoting an entire chapter to them later on. The regions of contact between processes of two or more nerve cells, across which an impulse passes, are called synapses. These technologies are concerned with being able to introduce neuronal tissue into the brain which can form functional synaptic connections with the rest of the brain and replace neuronal tissue which has been diseased or destroyed. These must be of an appropriate kind to reconstitute the normal synaptic circuitry of the brain. In addition, there are now known to be growth factors, referred to as neurotrophic growth factors, which are required in normal health to provide nutrients for the neurons in the brain. These may be artificially introduced into the brain to allow for the survival of neurons that would otherwise degenerate. They can also allow for injured or exogenously introduced neurons to form appropriate synaptic connections.

Rita Levi-Montalcini showed something extraordinary in a series of experiments, some of which were conducted as she hid from the Nazis in Turin during the Second World War. She obtained neurotrophic growth factors that allowed neurons which would otherwise degenerate to survive, and won the Nobel Prize for her discoveries.

Levi-Montalcini had a microtome which was used for cutting thin sections through the fixed embryos of birds. This, together with a microscope, enabled her to make fundamental discoveries concerning the development of neurons belonging to that part of the nervous system concerned with the control of internal organs, such as the heart. Levi-Montalcini showed that in this part of the nervous system, called the autonomic nervous system, neurons die quite normally during development so that there are more neurons present very early in life than in an adult. Many neurons in the peripheral nervous system are known as sympathetic neurons. Levi-Montalcini and Cohen purified the associated sympathetic nerve growth factor down to a single molecule and went on to show that neurons could be rescued from death if they were provided with this sympathetic nerve growth factor supplied by the targets with which autonomic neurons normally make synaptic connections.

Figure 1.8 shows a clump of sympathetic neuron cell bodies in a culture dish in the absence (upper figure) of this sympathetic nerve growth factor (NGF) compared with a clump in the presence of NGF (lower figure). Note that hundreds of axon processes emerge from the clump of neurons in the presence of NGF, indicating that these neurons are alive and growing.

Ten years ago my colleague Bogdan Dreher and I set out to see if what Levi-Montalcini had discovered for the peripheral nervous system (namely that autonomic neurons could be induced to survive if provided with the material from their normal targets such as cardiac muscle or smooth muscle) might also apply for neurons in the central nervous system. We first showed that retinal ganglion cell neurons, the nerve cells in the retina that send visual information from it to the brain along the optic nerve (depicted in Figure 1.9), normally die

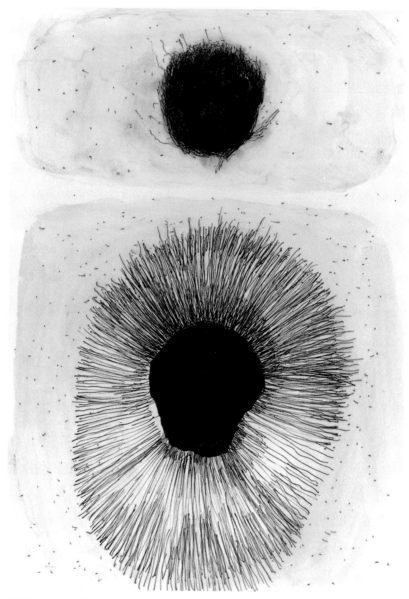

Figure 1.8   Sympathetic nerve growth factor effect on neurons.

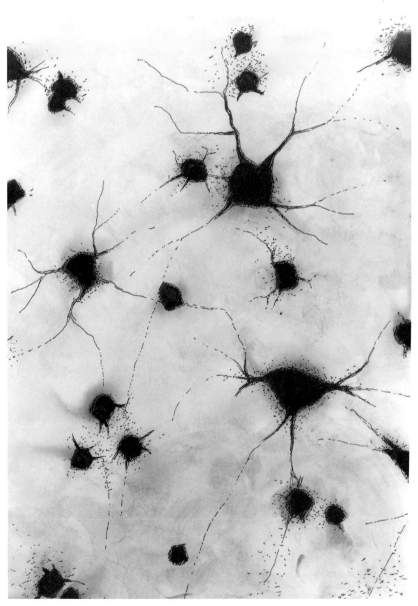

Figure 1.9   Retinal ganglion cells.

during development. Furthermore, these retinal ganglion cells could be induced to survive when provided with a nutrient neurotrophic molecule from their targets in the brain. Those parts of the brain are called the superior colliculus and the lateral geniculate nucleus. The neurons survived and sprouted nerve processes profusely in a tissue culture plate if provided with the neurotrophic factor, just as Levi-Montalcini had described for autonomic neurons. The difference was that in this case, the retinal neurons were supplied with a factor from the brain and not from muscle. This was probably the first indication that neurotrophic growth factors exist in the brain and not just in the peripheral nervous system, and that these growth factors can allow for the survival and profuse axon spouting of a central neuron such as a retinal ganglion neuron.

The question then arises as to whether neurons lying deeper in the brain, such as those belonging to the temporal lobe and to the hippocampus, also have growth factors which will support their survival. The neurotrophic factor which keeps retinal ganglion cells alive is not the same growth factor that Levi-Montalcini discovered in smooth and cardiac muscle that keeps autonomic neurons alive. To what extent can an injured temporal lobe or hippocampus be reconstituted by adding in a neurotrophic factor? The hippocampus, a relatively old and primitive type of cortex, abuts the temporal lobe. Depicted in Figure 1.10 is the limbic system of the brain which comprises primarily the inferior temporal lobe, hippocampus and amygdala. The hippocampus and amygdala lie on the medial surface of the brain

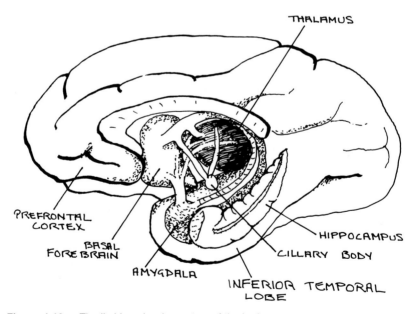

Figure 1.10   The limbic or border system of the brain.

rather than on its lateral aspect. Not only is the system most frequently involved in epileptic fits but it is also one of the first parts of the brain to degenerate in Alzheimer's disease. If a transverse section is cut through this region of the brain many different classes of neurons can be identified.

Figure 1.11 is such a section, isolated from the rest of the brain. The diagram shows the layout of the principal neuronal types (indicated as spheres with branching trees of processes called dendrites), together with their interconnections (indicated by the long thin processes, called axons, with arrows attached). The specific input, bringing information about all sensations from the cortex (such as sight, sound or smell), comes from the axons arising in the entorhinal cortex which synapse on the dendritic processes of granule cell neurons (particularly small neurons found throughout the brain) in the so called fascia dentata. These neurons themselves relay this transformed information to the pyramidal neurons in the CA3 region via their axons that form synapses on the CA3 neurons. These neurons are known to code for memories; they connect with each other through processes called recurrent collaterals (only two of which are shown). It is the excitability of these neurons, in part conferred upon them by these recurrent collaterals, that makes this region of the brain the most likely to trigger an epileptic seizure.

The CA3 neurons in turn project to the CA1 pyramidal neurons on which they synapse. These neurons also code for memories and their axons project to

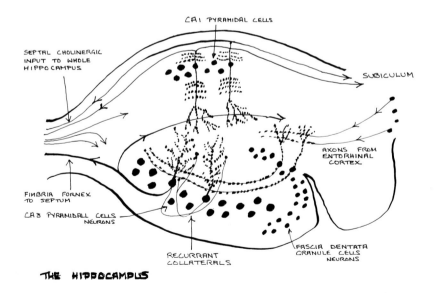

Figure 1.11   A section through the hippocampus, isolated from the rest of the brain.

the region called the subiculum in the old paleocortex and from there to the neocortex. Thus the trisynaptic pathway is from the entorhinal cortex to the fascia dentata to the CA3 pyramids to the CA1 pyramids back to the cortex. There is another less specific input (on the left) necessary for the normal functioning of the hippocampus and that comes from the structure called the septum. These axons synapse on all neurons in the hippocampus. They are the first neurons to degenerate in the brain of people suffering from Alzheimer's disease. This leads to the loss of memory formation that characterizes this disease. Surprisingly, these septal neurons are kept alive by Levi-Montalcini's growth factor; that is, the growth factor which keeps autonomic neurons alive in the peripheral nervous system. So if a neurotrophic factor is discovered that saves a certain class of neurons in the periphery it may well work on a whole class of neurons in the brain.

The synaptic connections within the circuit between the different types of neurons shown in Figure 1.11 have the special property of remembering over very long periods of time if they have been subjected to impulse traffic. Figure 1.12A shows a transverse section through the hippocampus, like that shown in Figure 1.11 except that stimulating electrodes have been placed on the nerves from the entorhinal cortex (called the perforant pathway axons) and a recording electrode has been placed in the dendritic layer of the granule cells where the

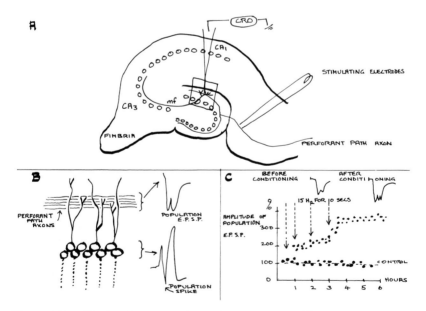

Figure 1.12 The functional synaptic circuits between the different types of neurons shown in Figure 1.11.

perforant pathway axons synapse. CRO indicates the cathode ray oscilloscope and mf indicates a mossy fibre. Mossy fibres are axons which project from granule cells in the dentate gyrus to CA3 pyramidal neurons. Figure 1.12B shows an enlargement of the boxed area in Figure 1.12A, with sample electrical recordings of the compound population spike (action potential). This population action potential occurs because many granule cells are firing action potentials in synchrony. Such synchronous firing occurs as a consequence of the stimulation (by the stimulating electrodes) of the perforant path axons that synapse on the granule cells. The population excitatory postsynaptic potential (epsp) is recorded from the synaptic regions between the perforant path axons and the granule cells on stimulation and gives a measure of the efficacy of synaptic transmission. Figure 1.12C graphs the relative amplitude of the population epsp against time and shows the results of stimulating the perforant pathway at a frequency of 15 Hz for ten seconds at the four times indicated by the arrows. If the perforant nerves are stimulated every few minutes with a single impulse before, during and after the 15 Hz stimulating periods, then the amplitude of the population epsp at each six minute period is shown to grow over three hours until it settles down to a size 300% that of the control (in which no stimulation occurred at 15 Hz). Two examples of this long-term potentiation of the population epsp are shown one before and one after the 15 Hz conditioning period. This enhanced efficacy of transmission through the synapses of over 300% is shown to last for three hours after the 15 Hz stimulation but may continue for days or months. The mechanisms responsible for this potentiation are required to retain a memory.

1.6 The restoration of memory processes by neural transplants into the brain

Rather than introducing growth factors into the brain in the hope of saving a certain class of neurons that are degenerating because of a neurological disease, it is possible to introduce embryonic neurons of the same class (or a set of neurons that have been genetically manipulated to be of a similar kind) as those that have degenerated.

Figure 1.13 shows the procedure for transplanting embryonic septal neurons from the hippocampal region of a fetus to the hippocampus of a mature animal with a degenerating septum. The septal region (black area) is first removed from the fetal brain and placed in a culture dish with enzymes that loosen the tissue into separate neurons. The partly separated neurons are then placed in a test tube and the isolation process taken further by rapidly shaking the neurons up and down in a pipette in the test tube. The completely dissociated neurons are next taken up into a syringe. The syringe needle is then located with great accuracy in the appropriate part of the hippocampus and the dissociated septal neurons injected into this area of the brain. They then reconstitute the normal circuitry, by making the correct synaptic connections, and restore the normal function, or at least almost the normal function, of that part of the brain which had degenerated. In this way, the normal elements of consciousness that are subserved by that part of the brain that has been injured or diseased are restored.

The synapses composing the internal circuitry of the hippocampus will not function to fire the neurons in the hippocampus unless there is another functional input from outside the hippocampus that just raises the efficacy of the hippocampal synapses so they are functional. This is provided by the septum. If the axons connecting the septum to the hippocampus are lesioned, then this cannot be provided. If the septal neurons are implanted into the hippocampus, they regrow their axons which again provide this increase in synaptic efficacy directly to the hippocampus in which they are embedded. What kind of tests can be carried out to see if (following degeneration of septal neurons) transplanted embryonic septal neurons can reconstitute the memory system in the hippocampus? Figure 1.14 shows one procedure that is used to determine this. It is called the alternating T-maze test. A rat is put at the end of one arm of the T-maze and is allowed to run to the other end where a choice is made of either turning left or right. In trial 1, food is placed on the right arm (at A in the diagram)

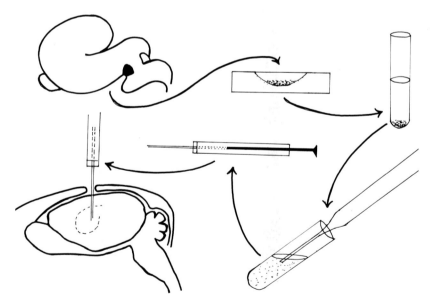

Figure 1.13   Procedure for transplanting embryonic septal neurons.

and the rat is forced to turn right because a trapdoor (door 2) prevents it from turning left. On trial 2, both doors are open so that the rat may now turn either left or right. However, before the rat makes the second trial, food is placed on the left arm of the maze (at B in the diagram). After this trial the rat is taken out of the maze and allowed to rest before another trial 1 – trial 2 session is performed. Altogether this is repeated about six times a day. The rat soon appears to determine

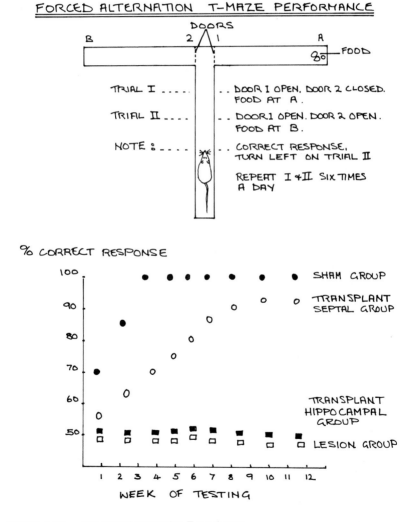

Figure 1.14 The forced alternation T-maze test.

what is going on. It turns right on the first trial and gets the food; on the second trial, when the food is put on the opposite arm, the rat immediately turns left and doesn't make the mistake of turning right. The graph in Figure 1.14 shows the rate at which the rat learns to make the correct choice on the second trial, namely to turn left. Following a sham operation in which the neocortex is exposed and then closed without any experimentation, the rat learns (within about three weeks) to always turn left to get the food on the second trial. If, however, there has been a lesion made in the septum, then the rat turns left on the second occasion at random frequency (namely on 50% of the occasions). It hasn't learnt at all; it cannot lay down the memory that it should turn left always on the second trial. But if healthy septal neurons from an embryonic rat are transplanted into the hippocampus of the adult rat with the lesioned septum (or of a rat whose septal neurons have degenerated because it has a form of senile dementia), then it only takes a matter of about two months before the rats can learn at the 90% level to turn left because normal hippocampal synaptic connections and functioning have been reconstituted by this septal transplant. The rat's memory system has been restored.

This experiment also establishes the specificity of the transplant. In contrast to the above transplant of septal neurons, if embryonic hippocampal neurons are transplanted into a rat hippocampus with a lesioned septum, then the rat performs no better than at chance level; that is, at 50%. The rat's memory system has not been restored.

Because the maze is symmetrical it might well be asked how it is that the rat can tell where it is in the T-maze: how can the rat tell its left from its right? The T-maze is set in a room which has lots of interesting objects in it including curtains, clocks, and pictures of rats as shown in Figure 1.15. These are essential for the rat to determine, when it is sitting at the start position on the T-maze, the geometry of the situation around it. The hippocampus calculates the spatial layout of the room from this information, allowing the rat to orient itself and so determine its left and its right.

In the T-maze experiment, an attempt was made to mimic the effects of senile dementia or stroke (that lead to degeneration of the septal neurons responsible for innervating the hippocampus) by introducing a lesion into the hippocampus. However, it is now possible to distinguish aged rats that suffer from a form of senile dementia (and so have a natural loss of septal neurons as a consequence of the dementia) from those that do not by using the Morris water tank (Figure 1.15).

A rat is placed at an arbitrary position in the water tank. The tank contains opaque water and a stand (shown in the cut-away of the tank wall) which is about one inch beneath the surface of the water; this is sufficiently deep for the rat not to see it when swimming in the tank so that it only becomes aware of the

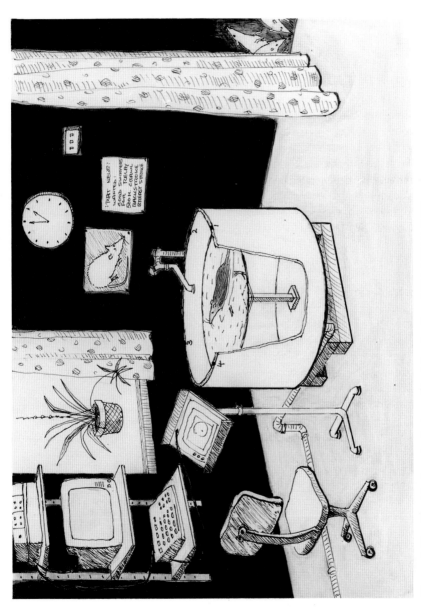

Figure 1.15 The Morris water tank used to determine the spatial memory of rats.

stand if its feet come in contact with it. Surrounding the tank are objects on the wall, such as curtains, clocks and pot plants (and a large picture of other rats) which allow the swimming rat to determine its orientation. The spatial location of the rat in the water tank is laid down as a spatial memory in the hippocampus and this allows a healthy rat to determine the position of the unseen stand with respect to the objects in the room, once its feet have come in contact with the stand. A television camera is placed above the water tank which allows the operator at the television-computer terminal shown to monitor the locus of the swimming pathway of the rat once it has been placed in the tank at an arbitrary position (Figure 1.16).

The locus of the movements of a young rat (about six months old) on the first trial on the first day are given in the first row and column of Figure 1.16. In this particular trial the rat's legs did not hit the stand beneath the water so the rat just swam around for a couple of minutes. But a few trials later, namely the fifth trial on the first day, the rat had learnt from the intermediate trials the position of the stand and so it swam immediately to the stand and sat down. Presumably, this

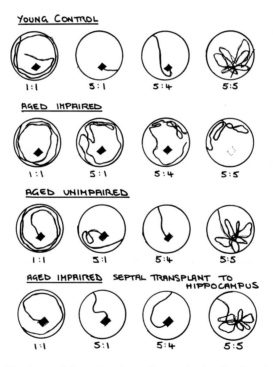

Figure 1.16　The locus of the swimming pathway of rats after they have been placed in the Morris water tank.

rat has formed a spatial memory of the position of the stand as a consequence of the normal functioning of its septo-hippocampal circuits. By the fourth trial on the fifth day the rat had no trouble at all; no matter where it entered the water, the rat immediately located the stand. On the fifth trial of the fifth day the following change was carried out: the stand was removed. The rat on entering the water tank at a random position swam around frantically trying to find the stand, and the locus of its movements were centred on the position where the stand had previously been. This technique can thus be used to pick out those rats that are suffering senile dementia from those that are not. In the second row of Figure 1.16 the results are shown for a 3½-year-old rat, about the equivalent of a human at eighty. It is apparent that even by the fifth trial on the fourth day, after the rat had bumped into the stand many times during the preceding days, it could still not locate the stand using spatial clues. This was an aged impaired rat; that is, it had senile dementia. All old rats do not get senile dementia any more than all old humans do. In the third row of Figure 1.16 the results are shown for another a 3½-year-old rat. By the fifth trial on the first day it had evidently formed a good spatial map of the whereabouts of the stand. On succeeding days it did as well as the young rat whose performance is shown in the first row.

Using this water tank technique, invented by Morris, it is possible to sort out rats suffering from senile dementia from those that are not. That is, it is possible to separate out those rats whose septo-hippocampal synaptic circuits are functioning from those in which these circuits are not. Furthermore, if an aged rat suffering from senile dementia has embryonic septal neurons introduced into its hippocampus, then after a month or so has elapsed (to allow the implanted neurons to make synaptic connections) the water tank test shows that the rat has recovered its ability to lay down spatial memories (see the last row in Figure 1.16). This approach clearly demonstrates the success of transplants of neurons from embryonic material which allow the reconstitution of damaged or degenerating synaptic circuits involved in the formation of spatial memory.

1.7 The idea of consciousness

We are now confronted with what I believe to be the third great area of mystery concerning natural knowledge. The first involved the discovery of quantum mechanics in 1926 by Schrodinger and Heisenberg. The second concerned the molecular basis of the genetic process and was to some extent begun by Schrodinger himself with his little book *What is Life?* which attracted physicists of high quality into biology. One of these was Francis Crick who after co-discovering the structure of DNA in 1952 went on to delineate the concept of the genetic code in specific ways which Schrodinger had only dreamed of in his little book. This helped lay the foundations of molecular biology that has come to dominate a good deal of biological science in the latter part of the twentieth century. The third challenge is the human brain and the origins of consciousness. How does consciousness develop and how did it evolve? A human embryo at six weeks is depicted in Figure 1.17. The hands are clearly visible with the just delineated fingers. The heart, with liver below, can be seen between the hands; the diaphragm separates the heart and liver. Most striking are the two halves of the cerebrum which can be seen through the transparent skin of the forehead above the developing eyes. How does this cerebrum develop so as to give rise to consciousness?

Questions concerning the evolution of human consciousness are considered further in Chapter 6. The growing knowledge of the brain and of how it gives rise to consciousness will offer insights into the means of ameliorating various neuronal diseases. In particular, those distortions of consciousness which arise as a consequence of vascular stroke and schizophrenia, as described in Chapter 5. By considering the neuronal concomitants of consciousness, this book is an attempt to illuminate this mysterious phenomenon, and so help us understand what we are.

1.8 References

1.   Moore, W. (1989) *Schrodinger*. Cambridge: Cambridge University Press.
2.   Schrodinger, E. (1944) *What is Life?* Cambridge: Cambridge University Press.
3.   Judson, H.F. (1979) *The Eighth Day of Creation*. New York: Simon and Schuster.
4.   Crick, F. (1988) *What Mad Pursuit*. London: Weidenfeld and Nicolson.
5.   Crick, F. (1994) *The Astonishing Hypothesis*. London: Simon & Schuster.

Figure 1.17 A human embryo at six weeks (15 mm long).

Figure 2.1   Rene Descartes (1596–1650).

CHAPTER 2   SYNTAX, SEMANTICS AND QUALIA IN CONSCIOUSNESS

2.1 The idea of brain and mind: Descartes and Kant

It was Descartes (Figure 2.1) who first posed the dualistic problem of the relationship between the brain, treated as an object for physical study, and consciousness. Kant (Figure 2.2) analyzed this problem further by distinguishing between sensory information, such as temperature and vision, that we receive through the excitation of different classes of sensory receptors and those activities that categorize these experiences as belonging to, for example, substances or to causal relations. To the first of these he gave the name Sensibilities and to the latter Categories of Understanding. Kant thought that the gathering of Sensibilities was most likely carried out by physical processes whereas the mind uses its Categories of Understanding to construct our awareness and comprehension of the physical world from the Sensibilities. The reception of sensory information and its early processing by the nervous system is generally agreed to be a physical procedure. This is thought to obey procedures that have a clear syntactical structure. Such a structure involves a systematic statement of the rules governing the formulas of a logical system, like those that determine the arrangement of words and phrases in sentences, or the organization of computer programs. On the other hand, our comprehension of the physical world requires an understanding of the meaning of signs and symbols, including things like sentences and words; this is the problem of semantics. At perhaps an even more complex level, there are also feelings and sensations that accompany our awareness of the world; the set of these that are associated with a particular object are called qualia.

The question arises as to whether the procedures of semantics or the development of qualia are such that they are carried out by physical means, like those generally agreed to be responsible for the syntactical mechanisms in the nervous

Figure 2.2   Immanuel Kant (1724–1804).

system involved in the generation of sensations. This chapter is concerned then with the problem of how syntax, semantics and qualia arise in the nervous system.

2.2 The neuronal basis of syntax: Allan Turing and David Marr

David Marr suggested that the syntactical structure of the information processing that is carried out by the Sensibilities can be divided into three levels. These are illustrated by means of the problem of what is called global stereopsis in which the visual system seeks to arrive at a three-dimensional reconstruction of an object that is being viewed by the retinas of both eyes. According to Marr, there are three levels at which this information processing task must be analyzed. In the first of these (involving computational theory), a definition is sought of the information processing problem whose solution is the goal of the computation. This involves characterization of the abstract properties of the computation to be carried out. Surprisingly, this is in general the most complex aspect of trying to seek a solution to an information processing task. In the case of global stereopsis it involves identifying the properties of the visible world that constrain the computational problem. For example, one of these might be that surfaces in the real world tend to lie in similar depth planes; smooth gradients in depth of the visual field are far more common than sudden changes or discontinuities. The second level of analysis is simpler and involves obtaining an algorithm; that is, a formal set of steps or procedures, which will carry out the computation. The third level involves physical realization of the algorithm. This may be implemented with the neurons of the brain or by a computer chip. According to Marr it does not matter whether the hardware is biological machinery or a computer as far as the computational or algorithmic levels of the information processing problem are concerned. It is known that the problem of global stereopsis is solved by neurons in the visual cortex at the back of the brain, but it could also be solved by a computer chip that was appropriately designed to compute the global stereopsis algorithm.

The distinction between the three levels of analysis of an information processing device (including the brain) can be illustrated by considering a very interesting property of retinal ganglion cells. In some species these neurons are directionally selective; that is, a ganglion cell fires impulses at a high rate when an object is moved past the overlying light-sensitive rod receptors that are connected to it in one direction (therefore called the preferred direction), yet when the object moves in the opposite direction the ganglion cell does not fire (and so this is called the null direction). An algorithm for this process has been developed and is shown in Figure 2.3.

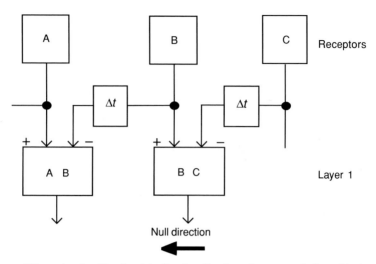

Figure 2.3 An algorithm for detecting the direction of movement of an object.

A, B and C are receptors that can sense the object, which moves over them in either the null direction (indicated by the arrow) or in the preferred direction. These receptors can each excite activity in the units immediately beneath them. Each box containing a Δt is a unit which if excited by the receptor connected to it will, after a delay of length Δt, prevent excitation of the adjacent unit in the null direction. When an object moves in the null direction, electrical activity from an excited receptor (say C) excites (+) a unit in the layer immediately beneath it while at the same time inhibiting (-) the next unit in the null direction; each receptor in turn, namely C, B, and A, carries out this process as the object moves over them. The delay units (shown as Δt) determine that the inhibitory process stops the excitatory activity from A and B moving through these gates if motion is in the null direction, but reaches the gates too late to produce such inhibition if motion is in the preferred direction.

The retina provides a biological means of implementing the directional selectivity algorithm given in Figure 2.3. The equivalence between the retinal neurons in Figure 2.4 and the elements that compose Figure 2.3 are: the rod receptors (R) are the receptors; the bipolar cells (B) are the layer one units; and the horizontal cells (H) are the Δt units. The output units, not shown in Figure 2.3, are the retinal ganglion cells (G). The direct excitatory synaptic pathway in the retina is from the rod receptors, through the bipolar cells, to reach the retinal ganglion cells. This excitatory pathway is modulated by the horizontal cells; they receive an excitatory synaptic input from the rod receptors, and conduct this laterally in the null direction through their long dendrites to inhibit the bipolar cells in adjacent regions. The wiring of the retinal neurons in this way prevents excitation

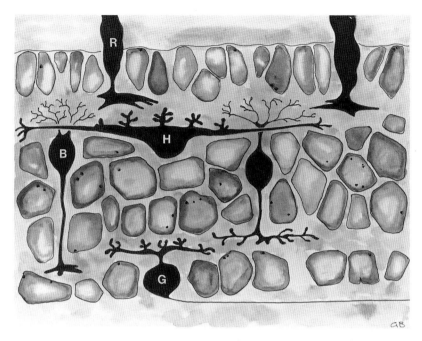

Figure 2.4   The retina provides a biological means of implementing the directional selectivity algorithm.

of the ganglion cells when an object moves in the null direction, but does not prevent such excitation when an object moves in the preferred direction. It is clear that the algorithm of Figure 2.3 can be implemented in either the wetware of neurons or the hardware of a silicon chip.

Although it has been stressed that the algorithm for global stereopsis as well as that for directional selectivity may be implemented using neurons or chips, they may also be carried out using a computer. It was Alan Turing (Figure 2.5) who first showed that given a particular algorithm, there could be built (in principle) a machine to solve it; this is often called the Turing machine. Turing went on to show that a Universal Machine could be built that could simulate any machine, so that this Machine could solve any algorithm. The present day computer is such a Machine. Programmed appropriately it can solve any algorithm. Suppose that all syntactical structures, from the rules governing the arrangement of words and phrases in sentences to those governing the logic followed by devices for directional selectivity, are algorithmic: it follows then that they can be implemented on a computer. The retina acts as one kind of Turing Machine when it solves the algorithm for directional selectivity.

Figure 2.5   Alan Turing (1912–1954).

The idea that different parts of the nervous system, such as the retina, have an objectively formal syntactical structure like that of a computer program has been challenged by John Searle[1]. He uses the word 'computational' in the context of solving the algorithmic problem rather than in Marr's sense of posing the problem and the constraints associated with it. His challenge goes like this: the computational state of the directional selectivity mechanism refers to the identification of the synapses in the system that are solving the algorithm for this problem. But it is we who make this selection, which is then not intrinsic to the retina like its temperature or mass, but assigned to it. Every neuron in the retina has thousands of synapses on it, so that it is impossible to look at the efficacy of each of these and determine that it is part of a directional mechanism algorithm. The algorithmic properties of the retina are assigned to it and are not intrinsic. The fact that algorithms can be implemented on different kinds of hardware, such as a retina or a silicon chip, simply shows that the computational processes of the algorithm are not intrinsic properties at all. They depend on interpretation from outside the system considered. That the computational patterns are carried out on a computer or in a nervous system does not explain how they work. Searle arrives at the important conclusion that syntax is an observer-relative notion. According to this view, what seems to be a relatively straightforward proposition—namely that different parts of the nervous system solve different sets of algorithms which are objectively posited in the system, like its temperature and mass—is erroneous. The proposition that a system has computational properties, intrinsic to it like those of its physical properties, is incorrect. Accordingly, syntax is observer relative.

2.3 The neuronal basis of semantics: Wittgenstein and Searle

Neuroscience has shown that different parts of the neocortex, the mantle of the brain, process different aspects of our experience of the world and of our reactions to those experiences. For example, the process of determining the shape of an object from the way it is shaded or the way that it moves is known to occur in an area of the neocortex that lies just in front of the primary visual cortex (compare Figure 2.6B with Figure 2.6A). Figure 2.6A shows the neocortex in a side view; several different regions are delineated that process different sensory modalities, such as smell (olfaction), sound (audition), touch (somatosensory) and sight (vision). Also shown is the area of the neocortex that is responsible for movement of the limbs (motor).

Figure 2.6B shows different regions of the neocortex that are thought by some to be involved in the solution of a number of different algorithms. Computing the movement of an object occurs in the parietal cortex whereas identification of objects or the reconstruction of the images of faces occurs in the inferior temporal lobe. The ability to process sounds that distinguish one word from another as well as their order occurs in Wernicke's speech area. Dennett[2] claims that each of these different areas of the neocortex is carrying out its task by solving an appropriate algorithm and that they are therefore acting as computers.

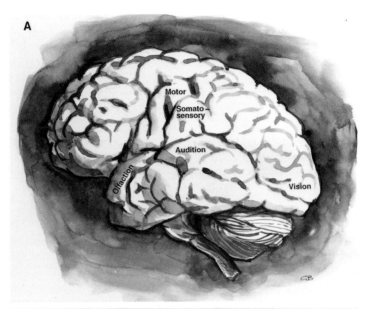

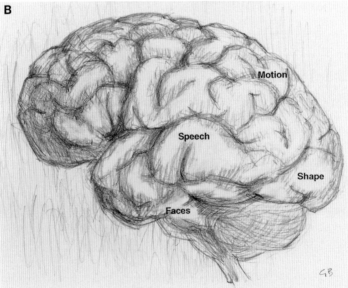

Figure 2.6 Possible algorithmic processes carried out in the neocortex.

Furthermore, the semantic content of these processes, their meaning, is the process itself. This is equivalent to saying that the meaning that we attach to the identity of an object is part of the algorithm that is carried out by the appropriate neurons in the inferior temporal lobe used for this purpose. In this case meaning is ascribed to the computational process that carries out the algorithm. Algorithms have been devised for carrying out the process of identifying objects from their shading (as biologically performed by the brain in the area anterior to the visual cortex). These are based on the kinds of information that this area is likely to receive from the visual cortex, and are framed in such a way as to offer a plausible description of what the neurons in this part of the brain do. Is it possible that such a computation also contains within it the meaning of the algorithm?

Searle[1] has a now famous argument that meaning is not embodied in algorithmic processes. Even if the idea is admitted that different parts of the brain possess intrinsic computational properties as a consequence of the way the neurons are connected (which Searle does not), the algorithmic process they carry out could have no meaning associated with it that is intrinsic. The meaning of the algorithmic process always lies outside the computational process. Searle's argument involves the idea of the Chinese Room (Figure 2.7).

Consider a number of people each of whom has been taught one component of the process involved in manipulating Chinese symbols according to instructions in English. This room receives Chinese texts through one delivery port, the same receptionist being used to receive each text. A text is then handed on to a person whose job it is to carry out the first process in manipulating the symbols according to the list of written instructions in English. This may involve certain rules of syntax being applied: for instance certain symbols in Chinese might be exchanged for certain other Chinese characters, each of which may be found in a separate basket. This person in turn passes this text on to the next person in the chain, and so on until the penultimate text is received by the last person who, after carrying out the last process on the text, delivers a completed manuscript to the exit port of the room. None of the persons in the room have any understanding of the processes carried out by any of the others, nor of the overall process. Yet an effective response to the original Chinese text may be achieved by this procedure even though the process is without meaning to those in the room. Now Searle suggests that the same argument holds if just one person carries out all these processes in the room; that is, manipulates symbols according to a set of rules (syntax). For instance, to those outside the room, some of the text presented may require questions on specific subjects to be answered. These are eventually supplied, and considered as answers to those outside the room, without having any meaning to those in the room. Searle's point is that an algorithm, such as that involved in manipulating the Chinese symbols, is

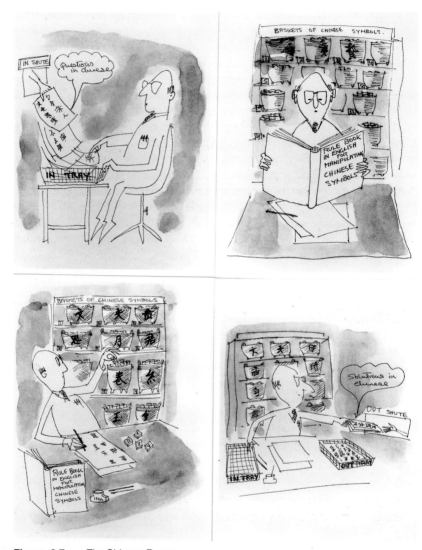

Figure 2.7   The Chinese Room.

computed without meaning being ascribed to any element of the process. This of course holds if we replace the persons in the Chinese Room with a neuronal network that carries out the computation. The Chinese Room only has meaning for the person who designed it in the first place; it does not arise from the workings of the room itself.

Wittgenstein (Figure 2.8) argued that meaning only arises as a consequence of some form of dialogue which presupposes a sociological setting. As there are an infinite number of such settings for each of us, and unique to each of us, semantics cannot be algorithmic. For example, when people talk about the color 'blue', they do not have an image of 'blue' in their brains to which they refer and which has somehow become imprinted there. Rather, it is the application of the word 'blue' to an object in a sociological setting that determines the character of the images which the person accepts as 'blue'. The meaning of this word is maintained by its public applications, not by reference to some template in the brain. Wittgenstein maintained that this also holds for words like 'pain' which do not apply to public objects: their meaning is maintained by sociological interactions involving the use of the word. It follows that semantics is embedded in an enormous variety of discourses and so cannot be algorithmic.

A further argument that meaning cannot be achieved by purely algorithmic processes is due to the mathematician Roger Penrose (see Chapter 7). His argument depends on the famous demonstration by Godel that any reasonably sophisticated system of axioms gives rise to statements that are obviously true but cannot be proved to be true within the particular system generated by the axioms. In this case one can appreciate that the statement is not formally provable within the system. As a consequence, Penrose claims that the concept of mathematical truth cannot be encapsulated in any formalistic scheme. If this is true, then no formal system of syntax can give rise to understanding, such as in the above case of identifying the truth of the statement. A computer or a neural network in our brains, which is following a formal set of rules, cannot then understand. This lies outside the system.

2.4 The neuronal basis of qualia: Edelman and Dennett

The attempt to incorporate qualia into an algorithmic procedure is much more difficult to conceive of than the possibility that syntactical or even semantic processes are algorithmic. The awareness of your mother in a room, sitting and reading, may involve qualia of movement, particular colors, the sound of her voice and the smell of her perfume. Such memories may be stored in the parietal cortex and the temporal lobes, where they have been built up over a huge number of experiences involving your mother in different sociological settings since the time you were born. It was a triumph of seventeenth century physics to rid qualia from objects for the purposes of describing their intrinsic properties, such as weight and temperature, which could be used as variables in different physical laws. Qualia are an essential ingredient of consciousness. The elimination of qualia for the purposes of reporting on those properties of an object that are useful in the formulation or carrying out of so-called physical laws is to

Figure 2.8   Ludwig Wittgenstein (1889–1951).

eliminate mind from nature. Wittgenstein pointed out that it is not possible to have knowledge of what we call qualia without consideration of the sociological setting in which this knowledge is obtained. In the words of Edelman[3], 'One cannot render obvious to a hypothetical qualia-free animal what qualia are by any linguistic description'.

It is possible, according to Dennett[2], to reconstruct qualia if they are merely 'dispositional states of the brain'. In this case qualia could be incorporated into algorithmic processes. For example, the brain may consider information about wine in the form of taste, color and smell. The disposition for a particular preference could be genetically built into the wiring of our neurons through evolutionary pressures, for example the color red may give a complex of conflicting emotional responses, arising from now largely redundant genetic information that red fruit is good and red snakes are bad. A more complex example, which attempts to face up to the criticisms of Wittgenstein, involves consideration of the qualia experienced when listening to a Bach fugue. The question is can the qualia associated with an auditory cortex listening to a fugue in Leipzig at the time of Bach (Leipziger auditory cortex in the upper panel of Figure 2.9) be reconstructed in the cortex of a contemporary listener in Boston (Bostonian auditory cortex in the lower panel of Figure 2.9). Dennett suggests that if a list is made of the differences between these periods and places, then it should be possible to interchange the qualia associated with the fugue between the two periods and places. However, Wittgenstein, Searle and Edelman have indicated that there are effectively an infinite number of such differences which are unique to an individual's experiences. Qualia cannot then be objectified in this way. They cannot be accomodated within a theory that draws up lists for each person of an entire life's experiences, embedded as this is in language and a unique sociology.

The neocortex is viewed by Dennett as a parallel computational processor of algorithms. The algorithms are each implemented in series by neurons in the neocortex. The location of some of the parallel processors associated with the implementation systems of audition, vision, movement and touch are indicated in Figure 2.10. Consciousness is then regarded as the computational process itself. According to Searle, however, these computations are not intrinsic properties of the brain but are observer dependent; that is, the experimentalist chooses the particular sets of synapses and neurons that he or she believes could carry out the computation which realizes the algorithm, which is very different from measuring intrinsic properties such as the temperature of the neurons. Furthermore, Searle uses the Chinese Room argument to show that meaning cannot arise from within an algorithmic process itself; this can only be ascribed to the person that designed the algorithm in the first place. If the brain is a parallel computational processor then, according to Searle, the meaning of these computations can only arise from outside the processors.

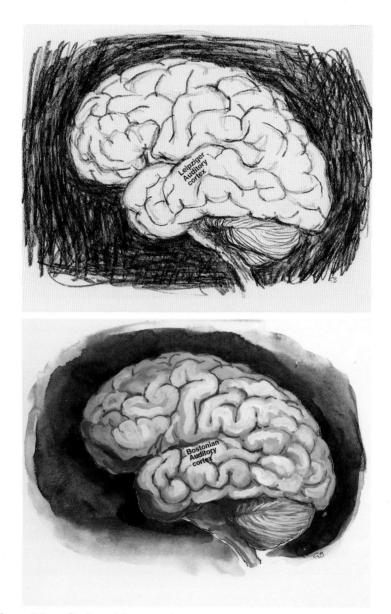

Figure 2.9   Qualia and the neocortex.

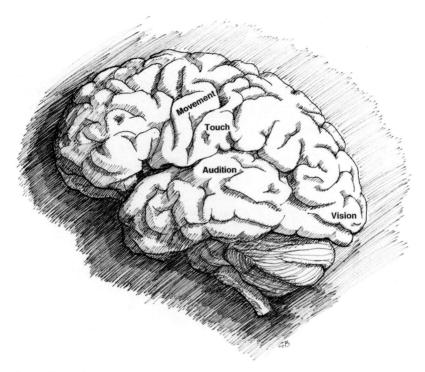

Figure 2.10   The location of some of the parallel processors associated with the implementation systems of audition, vision, movement and touch.

To what extent then has neuroscience illuminated our understanding of syntax, semantics and qualia? Neuroscientists certainly assume that the early processing of sensory information, like that giving directional selectivity in the retina, is syntactical. Furthermore, that algorithms can be found which carry out the appropriate computation, and that neurons exist which are connected up so as to carry out this computation. Yet it must be said that at this time Searle's[1] criticism of syntax as involving the execution of algorithms holds: no complete synaptic wiring diagram of a set of neurons subserving the implementation of an algorithm has been obtained. Nearly all wiring diagrams to this time assume that anatomical synaptic connections are functional; this assumption is very unlikely. In fact, most are not. Therefore, Searle's criticism concerning the subjective routine followed in the choice of synapses involved in integrative processes has not been refuted. As far as semantics is concerned, the claim that meaning arises from the computational process enacted in solving an algorithm may be refuted by the Chinese Room argument. Neuroscience at present has nothing to say about meaning. It may arise at levels of the neocortex that are so far removed from what we do know something about, namely the neural basis of early sensory processing, that it is as yet beyond neuroscience.

Neuroscience may have something to say about qualia. The neocortex contains different areas each specialized for dealing with different senses such as vision, audition and touch, as well as with distinguishing the sounds for different words (Wernicke's area) and with many other processes. There is some evidence to suggest that memories associated with a particular sensory modality are eventually stored in areas of the neocortex that are closely associated with that modality. For example, memory for a particular face is stored in the inferior temporal lobe, the area of the neocortex concerned with visually identifying objects. The possibility arises then that the qualia associated with your mother, such as her complexion, the sound of her voice and the smell of her perfume, are stored respectively in the visual cortex, the auditory cortex and the olfactory cortex. The many other qualia with which you identify your mother would likewise be stored in the appropriate part of the cortex, depending on the modality involved. In this way different components of the immense number of experiences you have participated in with her would be maintained in a distributed set of modules throughout the cortex. Your feelings and sensations about your mother would then require that these modules be activated in parallel. What the mechanism is for such activation and, more importantly, how such a distributed system could give rise to the holistic experience of a set of 'mother' qualia, is considered in the next chapter.

2.5 References

1.   Searle, J. (1992) *The Rediscovery of the Mind*. Boston: M.I.T. Press.
2.   Dennett, D.C. (1991) *Consciousness Explained*. New York: Penguin Press.
3.   Edelman, G.M. (1989) *Neural Darwinism, The Theory of Neuronal Group Selection*. Oxford: Oxford University Press.

CHAPTER 3 THE HOLISTIC NATURE OF CONSCIOUSNESS

When looking at an attractive scene, such as a garden, the viewer has an awareness of particular trees, shrubs and perhaps even flowers. It is extraordinary that just these named objects can be attended to amongst the enormous number of visual impressions that the retinas receive from the scene. For there is nothing to distinguish the photons which reach the retinas from, say, the bark on a tree and those from the shrubs surrounding it. What is it then that gives the experience of the tree as an holistic structure? What neural processes bind together its trunk, boughs and leaves into a single entity which is readily identifiable from the surrounding and sometimes partially enveloping shrubs? This is referred to as the binding problem and can clearly be considered at other levels of holistic experiences. For example, a breeze produces a rustle as it passes over the leaves of the tree. In this case there is consciousness of both the visual identity of the tree, involving the visual cortex, as well as the sound of the rustle of its leaves, involving the auditory cortex. How is the holistic experience generated when two quite different areas of the brain are involved? This chapter examines the solutions to the binding problem now offered by neuroscience, together with the problem of how it is that there can also be awareness of, or attention to, just particular solutions of the binding problem.

3.1 Neuronal groups and 40 Hz oscillations of neuronal activity

Thomas Young (Figure 3.1) conceived the idea that the neocortex might be thought of as being made up largely of neuronal groups. A neuronal group is a set of neurons in the neocortex which possesses a very large number of synapses that interconnect the individual neurons in the group. This large number of intrinsic connections is in contrast to the relatively sparse number of extrinsic synaptic connections from other nearby groups or from distant groups elsewhere in the

Figure 3.1   Thomas Young (1773–1829).

neocortex. Figure 3.2 depicts neuronal groups in the visual cortex, with extrinsic inputs from the retina through the lateral geniculate nucleus (LGN) in the thalamus.

A neuronal group can be identified functionally by means of electrical recordings. The first trace in Figure 3.3 shows recordings from an electrode placed within a cluster of neurons. The electrode records the summed electrical impulse firing from all the neurons in its vicinity, called a field potential. In the second trace another electrode is placed inside an individual neuron so that it only records the impulse firing from the impaled cell; it will be noted that there is heightened activity at the same time in the first and second traces. If the impulse firing of the

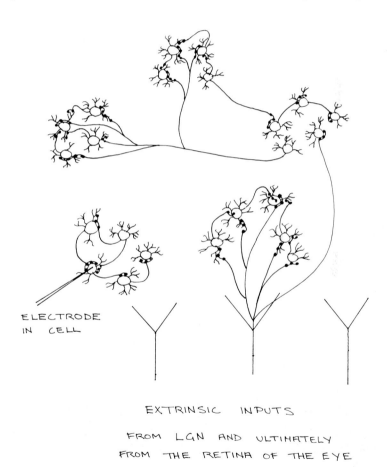

ELECTRODE
IN CELL

EXTRINSIC INPUTS

FROM LGN AND ULTIMATELY
FROM THE RETINA OF THE EYE

Figure 3.2   The synaptic connectivity of neurons that constitute neuronal groups.

**Multi unit activity (MUA)
and local field potential (LFP)**

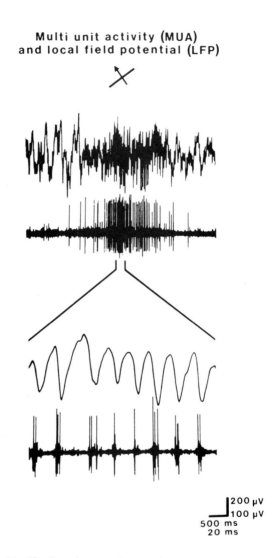

200 µV
100 µV
500 ms
20 ms

Figure 3.3   Identification of neuronal groups.[1]

impaled neuron is at the same frequency as that of the local field potential and occurs in phase with this field potential, then the recording is from a neuron within
a neuronal group. The third trace shows a small segment of the second trace on
an expanded time scale. It is now evident that the field potential is oscillating at
about 40 Hz; that is, the neurons in the vicinity of the electrode must be firing at
this frequency and in phase. The fact that the field potential is generated by a

neuronal group, of which the impaled neuron is a member, is confirmed by the fourth trace that shows the individual impulses in the impaled neuron which are firing in very fast bursts at intervals of about 25 ms; that is, 40 Hz.

Gerald Edelman[2] has argued at length that neuronal groups are to be considered as building blocks in the solution of binding problems. The very large number of axons connecting different parts of the neocortex are considered by him to subserve connections between neuronal groups. Such connections occur both between groups within a selected sensory modality as well as between sensory modalities. For example, the neuronal groups may be in the same area of the visual cortex. They may be in different areas of the visual cortex. They may even be in different hemispheres, such as the visual cortex in each hemisphere, connected by the very large number of axons, called the corpus callosum, that join the two hemispheres. Neuronal groups may also be connected across different sensory modalities, for example vision and touch. Some neuronal groups (indicated by clusters of dots) are shown on a side view of the brain in Figure 3.4. The very large number of horizontal synaptic connections between the myriads of neuronal groups in the neocortex are not shown. Sets of groups are shown, however, in the visual cortex, in the auditory cortex, and in the somatosensory cortex.

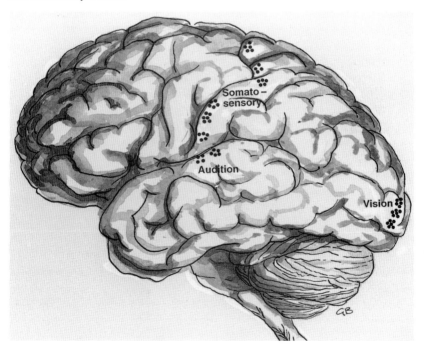

Figure 3.4 Some neuronal groups on a side view of the brain.

3.2 Neuronal groups participate in the solution of the binding problem in the visual cortex

Wolfgang Singer[1] has provided evidence that there is a dynamic coupling of appropriate neuronal groups in the neocortex with the solution of a binding problem. Consider, for example, two vertically oriented light bars moving at the same speed in the same direction past the eyes. Despite the fact that the bars are sufficiently far apart to be seen by two quite different parts of the retina which project to two distinct neuronal groups in the visual cortex, there is a tendency for these bars to appear as a single object.

The photomicrograph in Figure 3.5 shows the position of two such neuronal groups in the visual cortex of a cat. The bars are sufficiently far apart to be seen by two quite different parts of the retina which project to two neuronal groups in the visual cortex that are seven millimeters apart (as indicated on the photomicrograph of the surface of the visual cortex by the white arrows). The black areas in this flat mount of the surface of the cortex indicate columns of neuronal groups, of which only the tops are shown, that are particularly responsive to vertical light contours. Microelectrodes are placed in the vicinity of these two neuronal groups and the recordings made are shown in Figure 3.5A. The average impulse firing of the neurons in each of the groups (as shown by the field potentials) is oscillatory. Figure 3.5B shows on an expanded time scale that the oscillation of both of the groups is at 40 Hz and that they are in phase, despite the fact that they are seven millimeters apart. No such coupled firing would be expected for neuronal groups at such a distance, and because of this coupling it is likely that these two light bars would appear as a single object to the cat. The experience that the two light bars are one object is correlated with the fact that the neuronal groups in the visual cortex which are independently excited by the image on the retina of just one of the bars are now joined in a 'dynamic way'; that is, by a common frequency and phase of neuronal firing. This is an example of the transient excitatory coupling of two neuronal groups within the same area of the neocortex, in this case the visual area.

Singer[1] has also shown that there is interhemispheric synchronization of activity in the visual cortex when a binding problem is being solved for a visual object.

Suppose a single light bar is sufficient to stimulate three different neuronal groups, about one millimeter apart, in the visual cortex of one hemisphere. Electrodes are positioned in each of the groups. Numbered from 1–6 in Figure 3.6A are three such electrodes in area 17 of each hemisphere of the visual cortex of a cat. The synchronization of the impulse firing and the phase of this firing, as measured by the different electrodes, can be shown by means of what is called a cross-correlogram. Figure 3.6B depicts intrahemisphere cross-correlograms for the field recordings of pairs of neuronal group activity indicated by the electrode positions 1–3, 2–3, 4–6 and 5–6. If a periodic pattern is discernible in the cross-coòrelogram, then this indicates that the signals are correlated and gives information as to the common frequency and phase in the correlation. The cross-correlograms for the intrahemisphere recordings show a strong oscillatory modulation in the same frequency range of about 40 Hz, even though the electrodes may be separated by as much as two millimeters (Figure 3.6B). The cross-correlograms for interhemisphere recordings show surprisingly similar correlations (not depicted in Figure 3.6) to those in Figure 3.6B, indicating that both hemispheres participate in the solution of the binding problem for the single white bar.

This is not the case if the group of axons that join the two hemispheres (the so-called corpus callosum) is cut. The cross-correlogram for recordings from the

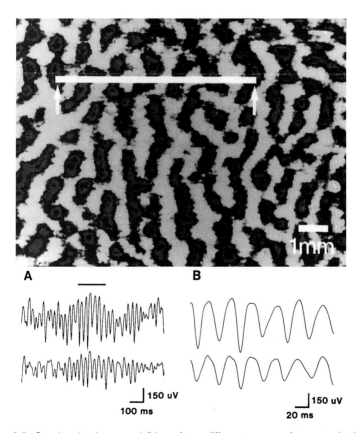

Figure 3.5 Synchronized neuronal firing of two different groups of neurons in the visual cortex of a cat (area 17) during the observation of two vertically oriented light bars moving with the same speed and in the same direction.[3]

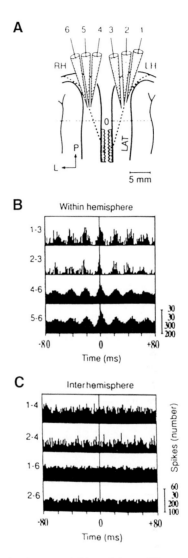

Figure 3.6   Synchronized neuronal firing of three different groups of neurons in each hemisphere of the visual cortex of a cat.[4]

two hemispheres is now devoid of any periodic pattern and is flat, indicating that the firing of neuronal groups due to the light bar in each of the hemispheres is no longer correlated (Figure 3.6C). The corpus callosum therefore mediates the dynamic connections between the two hemispheres that most likely participate in the solution of the binding problem.

3.3 The dynamic coupling of neuronal groups

Several plausible models are now available that explain how dynamic coupling between neuronal groups might arise during solution of the binding problem by the neocortex[5]. For simplicity, a neuronal group may be modelled as a single oscillatory element. This element oscillates as a consequence of the coupling of an excitatory unit (one that only forms excitatory connections) with an inhibitory unit (one that only forms inhibitory connections); each of the connections possesses delays built into them. An input unit allows for extrinsic connections to the excitatory unit. Activity of a single oscillatory element depends on the delay time in the connections, on the level of input from the extrinsic unit, and on the strength of the coupling which each connection makes with a unit in the element. These elements are referred to as delayed non-linear oscillators.

Figure 3.7 depicts such a model. The symbols on the excitatory (+) and inhibitory (−) units and their connections refer to the following properties which these have: $x_e(t)$ indicates the activity of the excitatory unit and $x_i(t)$ indicates the activity of the inhibitory unit; $F(x)$ gives the output function of a unit; w_{ei} gives the strength of the synapse from the excitatory unit to the inhibitory unit and w_{ie} gives the strength of the synapse from the inhibitory unit to the excitatory unit; T_{ei} gives the time delay built into the connection between the excitatory unit and the inhibitory unit and T_{ie} the time delay between the inhibitory unit and the excitatory unit; ie(t) indicates the extrinsic input to the element. This element will oscillate at a characteristic frequency determined by the values of the parameters.

Synchronization between oscillatory elements in a layer, equivalent to synchronizing the activity between neuronal groups across a part of the neocortex, is achieved by coupling the excitatory unit of one element with the inhibitory unit of another (Figure 3.8). Consider, for example, a set of elements laid out in a rectangular matrix of fourteen by seven elements. At first, all the inhibitory connections between the elements have been made but none of the excitatory connections; in addition, the oscillatory activity of each element (for which T is the period length of oscillation of an isolated element) is desynchronized by a noisy extrinsic input.

Figure 3.9 shows the activity of twenty delayed non-linear oscillator units (each representing a neuronal group) chosen at random from the two-dimensional matrix of fourteen by seven coupled elements (which may be considered to be representing a large array of neuronal groups in, for example, the visual cortex). This graph shows that if the excitatory connection between elements is not connected and there is a high noise input from an extrinsic element, then activity of the twenty chosen elements is highly desynchronized (the desynchronized time is taken from −5T to 0). At time 0 these connections were made and all twenty elements started to oscillate together in phase.

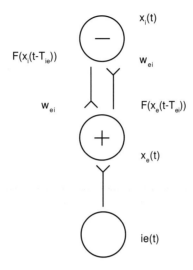

Figure 3.7   Modelling of a neuronal group by a delayed non-linear oscillator.[5]

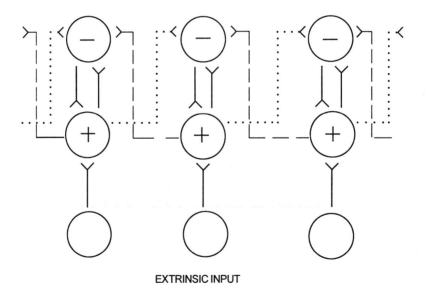

EXTRINSIC INPUT

Figure 3.8   Modelling of the horizontal connections between neuronal groups by coupled delayed non-linear oscillators.[5]

This effect can also be illustrated by means of an activity-phase map like that shown in Figure 3.10: each circle represents an element in the fourteen by seven matrix, with the diameter of the circle proportional to the activity of the element and the degree of shading proportional to the oscillation phase. When all the nearest neighbor excitatory couplings are made between nearest neighbor elements (of which there are eight for each element), then all elements fire with maximum activity and all are in phase (upper matrix). When these connections are broken, the activity of many of the elements falls to near zero and those that remain active are generally out of phase (lower matrix). This example shows how the appropriate coupling of inherently oscillatory elements leads to maximizing their activity and brings them all into phase, a necessary condition for a correct model of the dynamic coupling of neuronal groups.

How then do these delayed non-linear oscillators provide insights into the solution of the binding problem? Consider a two-dimensional array of such elements, coupled together by excitatory connections, each representing a neuronal group. These groups might be taken to analyze in the visual cortex a small part of the visual scene projecting onto a small part of the retina. The effect of coherency in the stimulation of different members of these elements, giving

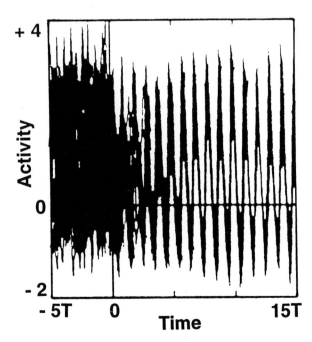

Figure 3.9   The activity of twenty delayed non-linear oscillator units.[5]

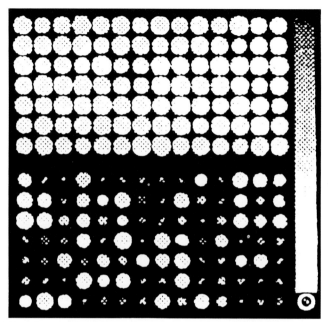

Figure 3.10   An activity phase map of all fourteen by seven non-linear oscillator elements (or neuronal groups).[5]

rise in a real cortex to synchronization of their activity associated with the solution of a binding problem, is considered for two short light bars.

These two bars are taken to provide extrinsic excitation of the elements as shown in Figures 3.11A and B; note that the light bars provide excitation of groups which are four elements apart in A and only two elements apart in B. Another example is provided by a continuous light bar equal in length to the two short bars (Figure 3.11C). Figures 3.11D, E and F show the cross-correlations for the numbered elements in Figures 3.11A, B and C respectively. The dashed line shows the cross-correlation within the stimulus bar segments (numbers 1–2 and 3–4 in Figures 3.11 A, B and C) and the continuous line shows the cross-correlations between the bar segments (numbers 2–3). The cross-correlations between bar segments are equivalent to stimulus coherency; that is, the two bar segments are bound together as one object. This occurs most strongly for the light bar segments when they are reasonably close together, as in Figure 3.11B, and is shown by the large size of the cross-correlation between elements 2–3 in Figure 11E. Note the fact that this cross-correlation is strong despite the fact that the two excited groups are separated by two non-excited elements. When the bar segments are relatively far apart, as in Figure 3.11A, the cross-correlation between 2–3 is very small (Figure 3.11D). The cross-correlation for the numbers within each of the short bar segments and within the long bar segment is always high; that is, for numbers 1–2 and 3–4. This example shows that correlated impulse firing of neuronal groups that are separated in space may be expected if there is some form of stimulus coherency.

3.4 Paying attention

Although mechanisms of the kind described above may be involved in the solution of the binding problem, they do not necessarily provide an explanation of how it is possible to attend to, or be aware of, a particular object in the visual field. There may be a number of different objects in the visual field, with associated images falling on the retina. The visual neocortex may then solve the binding problem for each of these, but what then is the mechanism involved in the process of attending to, or being aware of, only one of these objects? The mechanism of selective attention is well illustrated by the experiment shown in Figure 3.12. Recordings are made of the electrical activity of a neuron in the part

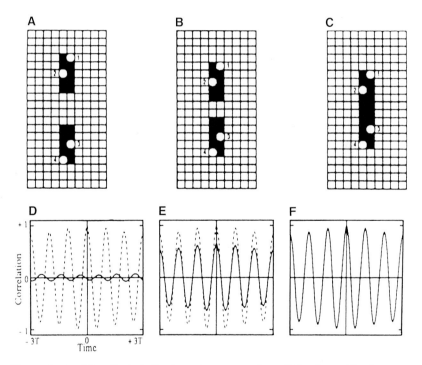

Figure 3.11   The effect of stimulus coherency on the activity of a matrix of non-linear oscillator units (representing coupled neuronal groups in the visual cortex).[5]

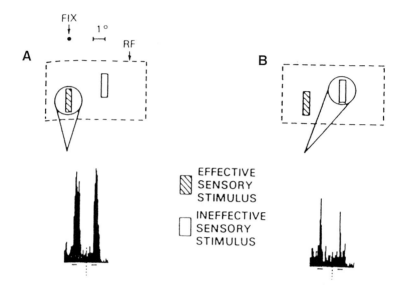

Figure 3.12   The effect of selective attention on the firing of a neuron in the visual cortex of a monkey.[6]

of the monkey neocortex that subserves vision. The neuron fires impulses when the image of an object falls over any part of a fairly large area of the retina, indicated by the large dotted rectangle in the figure; this area is called the receptive field of the neuron. A red bar (hatched bar in Figure 3.12) presented anywhere over the area of this receptive field produced vigorous firing of the neuron, whereas a green bar (open bar in Figure 3.12) failed to excite the neuron at all. The animal was then taught to attend to different locations within the receptive field, using an appropriate reward procedure. In the first experiment, the monkey attended to the location of the red bar (the attention being indicated in Figure 3.12 by a searchlight), whilst a green bar was also introduced into the receptive field; under these conditions the neuron fired vigorously, as shown in the lower part of Figure 3.12A. In the second experiment, the monkey was taught to attend to the location of the green bar, so ignoring the location of the red bar that was still maintained within the receptive field (Figure 3.12B). In this case, the neuron fired at a much lower rate (lower part of Figure 3.12B) even though the red bar was still within the receptive field of the neuron and might then be expected to give rise to vigorous firing of the neuron. It is clear that lack of attention to the location of the red bar greatly decreased the rate of firing of the neuron. This experiment shows that neuronal mechanisms that are responsible for attention must be engaged, possibly in addition to those that solve the binding problem, in order for a neuron that is involved in high-order visual processing to fire impulses at an optimal rate.

We have seen that reciprocal connections between neuronal groups in the neocortex can lead to 40 Hz synchronized oscillations of action potential firing in a particular set of groups that are solving a binding problem. What then are the mechanisms that determine which particular solutions of the binding problem will be attended to and not others? This is equivalent to asking (in the case of visual phenomena): what objects in the visual field for which the binding problem has been solved in the neocortex will be allowed to reach consciousness?

Francis Crick[7] has suggested that the attentional mechanism for this process is centered in the thalamus. The sensory inputs to the neocortex arise from the thalamus (Figure 3.13). The principal neurons of the thalamus in turn receive connections from such sources of sensory information as the retina. The principal neurons in the thalamus are surrounded by a concentric layer of neurons called the reticular complex; this receives connections from the neocortex as well as from axons that leave the thalamus on their way to the neocortex. Most importantly, these reticular neurons make connections with the principal neurons of the thalamus that are only inhibitory. Within the thalamus itself there is another set of neurons, called the pulvinar, which also receives connections from the neocortex as well as projecting extensively to the neocortex. These pulvinar neurons, like those of the reticular complex, make inhibitory connections with the principal neurons of the thalamus. Crick has argued that if there exists consciousness of a particular aspect of the visual field then an attentional mechanism is operating that involves the reticular complex and the pulvinar.

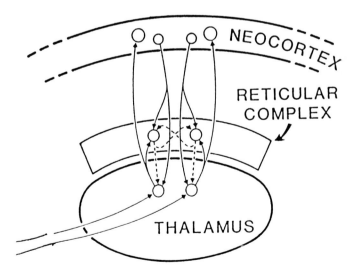

Figure 3.13 Sensory inputs to the neocortex, such as vision, must project through the thalamus (with the exception of the sense of smell).[7]

Neurons in these structures depress the sensory input from the thalamus to the neocortex that is not relevant to solving the binding problem associated with this particular aspect. The neuronal groups in the visual neocortex that are solving this problem are therefore given an excitatory advantage over those solving other binding problems.

The discovery of principal neurons in the thalamus that fire impulses in the frequency range from 30 Hz to 40 Hz, and which possess an underlying rhythmicity of about this frequency, does much to support the idea that the thalamus participates in the generation of 40 Hz oscillations in the neocortex.

The electrical recordings from two principal neurons in the thalamus, which illustrate that these have intrinsic impulse firing rates of about 40 Hz, are given in Figure 3.14. There is an oscillation of the baseline potential in these neurons that occurs at about 40 Hz. (Each oscillation is called a prepotential, one of which is indicated in the figure by an arrow.) These oscillations are sometimes large enough to fire impulses, two of which are indicated by an asterisk. These results suggest that at least part of the 40 Hz oscillations recorded in the neocortex arise from this kind of oscillatory activity in the thalamus.

Further support for the idea comes from studies on patients with damage to the reticular complex or the pulvinar on one side of their brain. Defects in attention are readily discerned in such patients by, for example, their response time to a visual target which is presented at some random time interval after a cue.

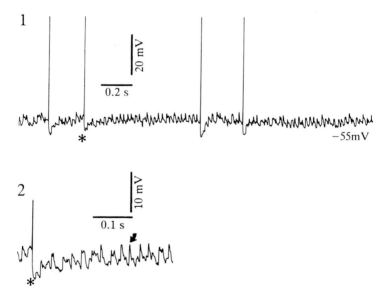

Figure 3.14   Electrical recordings from two principal neurons in the thalamus.[8]

Figure 3.15 shows the results of such an experiment: the vertical axis on the graph shows patients' reaction times to respond to visual targets, and the horizontal axis gives the intervals between presentation of the cues and that of the targets. If the cue–target combination was presented to the part of the visual field subserved by the damaged thalamus, the reaction time was always about one-fifth of a second slower than when the cue–target combination was presented to the part of the visual field subserved by the normal thalamus on the other side of the brain. This was the case no matter what the interval between presentation of the cue and that of the target. Attentional mechanisms therefore appear to involve the reticular complex and the pulvinar of the thalamus in humans and other animals, as Crick first suggested. Crick called his theory of attention the searchlight hypothesis[7], as in a sense the reticular complex and the pulvinar promote only a small proportion of the activity that reaches the thalamus on its way to the neocortex, and this increased activity can then be likened to a searchlight that lights up a part of the neocortex. The patches of neocortex that are solving binding problems for the objects attended to will then possess much greater activity than other patches. A final set of coherent oscillations then emerges only for those neuronal groups which the searchlight has sought out. These then constitute consciousness of the attended objects.

If the reticular complex and pulvinar of the thalamus participate in attentional mechanisms in this way, then what determines which particular groups of neurons in these regions of the thalamus are active? Put another way, what determines that we attend to a particular object over another in our visual field, this attending

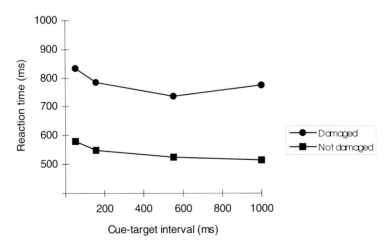

Figure 3.15 Evidence that the thalamus contains neurons that are necessary for attentional behavior.

process requiring the enhanced activity of just a particular set of neurons in the reticular complex and the pulvinar of the thalamus?

The brain stem, at the top of the spinal cord, contains three groups of neurons that project to the reticular complex, and which have a powerful modulating effect on the neurons there. Figure 3.16 is a medial view of the brain showing the sets of brain stem neurons in question: the raphe nucleus (RN) and the locus coeruleus (LC); the third set, the peribrachial nucleus, is not shown. The raphe neurons secrete the transmitter substance serotonin from their synaptic terminals in the thalamus, the locus coeruleus neurons secrete the transmitter norepinephrine and the peribrachial neurons secrete the transmitter acetylcholine.

Experimental evidence which shows that these pathways mediate the firing of the 40 Hz oscillations in the cortex associated with the solution of the binding problem and with attentional mechanisms is given in Figure 3.17. Electrodes were placed over a part of the neocortex in order to record the electroencephalo-gram (EEG) depicted; this gave a measure of the extent to which neurons were firing impulses at different frequencies in the neocortex. The amplitude of the waveform indicates the number of neurons which were firing at the frequency rate marked on the horizontal axis.

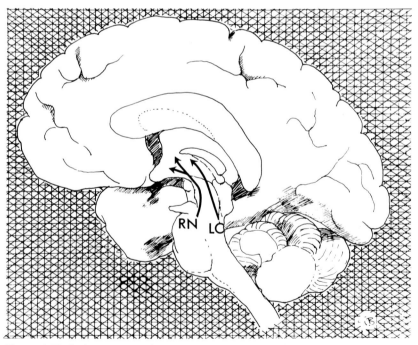

Figure 3.16 Brain stem sets of neurons that project to the reticular complex and the pulvinar of the thalamus.

The left hand-set of records in Figure 3.17 shows twenty separate power spectra of EEG recordings taken from the neocortex (over a part called the suprasylvian gyrus). The fact that the maximum size waveform occurs at around 40 Hz shows that most of the impulse firing cells were discharging at about this frequency. The first ten records (from −10 to 0) were controls, whereas the next ten records (from 0 to 10) were taken during stimulation of the brain stem parabrachial neurons. The increase in the waveform indicates that such stimulation greatly increased the number of neocortex neurons firing at 40 Hz.

The other twenty recordings depicted were taken under the same conditions as the left-hand set of records with the exception that the action of the transmitter released at synapses in the thalamus from the brain stem parabrachial neurons (acetylcholine) was blocked with the drug scopolamine. In this case, the waveforms corresponding to 40 Hz oscillation in the control condition (from −10 to 0) were virtually wiped out (indicating that the number of neocortical neurons firing at 40 Hz had been greatly reduced). Thus stimulation of the parabrachial neurons had no effect on the number of neocortical neurons firing at 40 Hz.

The process in which a set of subsidiary synapses either allows a main set of synapses to function, and transmit successfully, or not, is known as the gating process. When analogous observations to those made in relation to Figure 3.17 are made for the other brain stem groups, it is confirmed that the control of the reticular complex and the pulvinar by the brain stem determines the gating process on the principal neurons of the thalamus. The brain stem then appears to

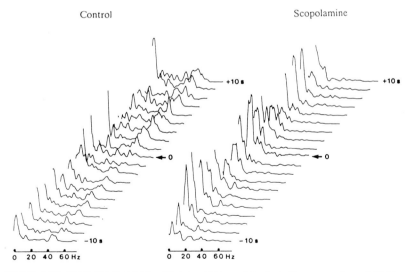

Figure 3.17 Evidence that brain stem neurons participate in the generation of 40 Hz oscillatory impulse firing in the neocortex.[8]

have a major role in determining that the neocortex will possess neurons that are firing at 40 Hz. It would seem natural that this should be the case, as brain stem neurons are also involved in controlling the sleep–wake cycle, namely in determining the general extent of arousal of the neocortex. Although it is clear that the brain stem has a major excitatory effect on the gating mechanism in the thalamus, this is not to say that our attention is determined by the brain stem. There are more nerves projecting from the neocortex to the thalamus than from the thalamus to the neocortex (note that Figure 3.13 does not show this quantitatively). It is very likely that the neocortical projections to the reticular complex and the pulvinar play a major role in determining which neurons in these areas will be activated, and that the brain stem input ensures that the neurons so chosen are excited.

3.5 Consciousness as a global map of neuronal groups

Crick's[7] idea then, is that the thalamic reticular complex and the pulvinar interact with the brain stem and neocortical mechanisms to reach a salient decision as to which neuronal groups that are active will be brought into consciousness by the spotlight of attention (Figure 3.18). Edelman[2] has a rather similar notion of how consciousness arises. He refers to the neuronal groups that are participating in the synchronous 40 Hz firing as constituting a 'global map' which at any particular time provides the content of what he refers to as primary consciousness. This global map is chosen by interaction with the emotional centers in the hypothalamus and brain stem through the hippocampus; these in turn interact with the frontal lobes, parietal cortex and temporal lobes to draw on the memory there previously laid down by the emotional emphasis given to previous global maps (Figure 3.18). These parts of the neocortex then interact with the global map neuronal groups to settle on the final global map. It is this map that constitutes consciousness. Edelman's mechanism of choosing the final global map is in principle the same as that suggested by Crick: neuronal groups in the neocortex solve different binding problems but only a particular set of these is chosen by the spotlight of attention to form the final global map. The difference between Crick and Edelman is in the details of what parts of the brain constitute the spotlight of attention. The important claim by both Crick and Edelman is, of course, that the final global map neuronal groups at any particular time constitute consciousness at that time.

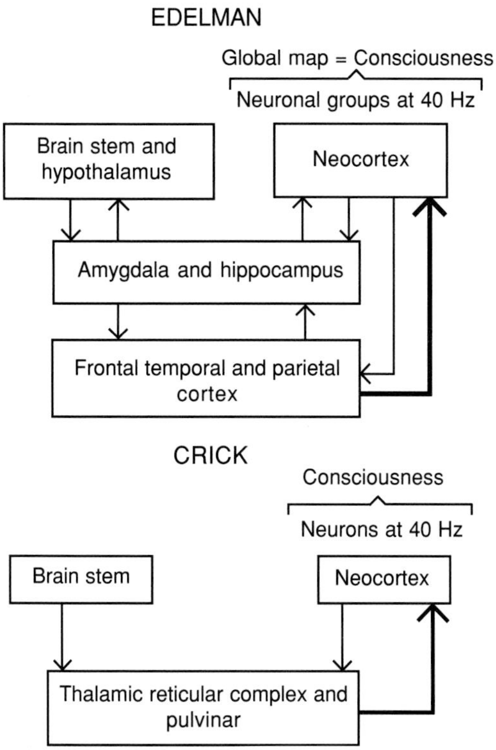

Figure 3.18   Comparison of the ideas of Edelman and Crick for generating consciousness.

3.6 References

1.   Singer, W. (1991) Response synchronization of cortical neurons: an epiphenomenon or a solution to the binding problem? *International Brain Research Organization News*, **19**, 6–7.

2.   Edelman, G.M. (1989) *Neural Darwinism, The Theory of Neuronal Group Selection*. Oxford: Oxford University Press.

3.   Gray, C.M. and Singer, W. (1989) Stimulus specific neuronal oscillations in orientation columns of cat visual cortex. *Proceedings of the National Academy of Science of USA*, **86**, 1698–1702.

4.   Engel, A.K., Konig, P., Kreiter, A.K. and Singer, W. (1991) Interhemispheric synchronization of oscillatory neuronal responses in cat visual cortex. *Science*, **252**, 1177–1179.

5.   Konig, P. and Schillen, T.B. (1991) Stimulus dependent assembly formation of oscillatory responses: Synchronization. *Neural Computation*, **3**, 155–166.

6.   Moran, J. and Desimone, R. (1985) Selective attention gates visual processing in the extrastriate cortex. *Science*, **229**, 782.
7.   Crick, F. (1984) Function of the thalamic reticular complex: the searchlight hypothesis. *Proceedings of the National Academy of Science of USA*, **81**, 4586–4590.
8.   Steriade, M., Curro Dossi, R., Pare, D. and Oakson, G. (1991) Fast oscillations (20–40 Hz) in the thalamocortical systems and their potentiation by mesopontine cholinergic nuclei in the cat. *Proceedings of the National Academy of Science of USA*, **88**, 4396–4400.

CHAPTER 4   THE CONSCIOUSNESS OF MUSCULAR EFFORT AND MOVEMENT

4.1 The simplest nervous system

How is it that the heaviness of an object held in the hand reaches consciousness? Understanding this problem involves a study of the mechanisms that control movement and sensation in the limbs, and begins with a consideration of the evolution of the nervous system which may have started 1000 million years ago with the sponges, or just 650 million years ago with polyps like jellyfish and corals.

The phylum Porifera, or sponges, are very simple multicellular parasites. They do not possess a nervous system and it is controversial as to whether they have any neurons. Figure 4.1A shows the simple motor reactions of the freshwater sponge *Ephydrata*, with the mouth chimney changing its form as water is drawn into the body through small pores and passed out through the mouth. Figure 4.1B shows some of the cell types that stain with silver in the sponge *Sycon raphanus*. The outer surface is connected to the inner surface by two cells with long and thin processes that may be primitive nerve cells; the cells on the inner surface are collar cells (called choanocytes).

With the evolution of the coelenterates, such as hydra, jellyfish and corals, the identification of nerve cells and muscles is unequivocal. Jellyfish have two layers of cells with a jelly-like substance separating them, giving the animal some rigidity (Figure 4.2A). The two different kinds of neuron networks (nerve nets) for the control of swimming, tentacle position and feeding that are present in *Aurells aurata* are shown in Figure 4.2B. One is composed of neurons that possess two axons (bipolar neurons), and these are prominent in relation to the radial and circular muscles that are exposed in this drawing; the other is composed of neurons with more than two processes (multipolar neurons), and these are shown in relation to the gastric cavity. Both bipolar and multipolar

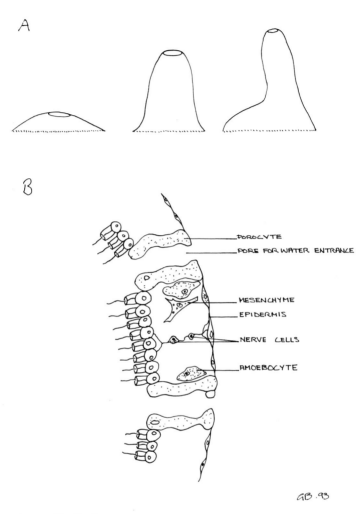

A

B

POROCYTE

PORE FOR WATER ENTRANCE

MESENCHYME

EPIDERMIS

NERVE CELLS

AMOEBOCYTE

GB .93

Figure 4.1   The Porifera.

neurons from each nerve net are apposed to each other in collections of neurons called ganglia (which serve as integrating centers), as shown in Figure 4.2C. Here the input to the bipolar cells associated with the muscle is transferred to the multipolar neurons associated with the gastric cavity. This collection of neurons into a ganglion for the purposes of neural integration occurs for the first time in evolution in the Coelenterates.

A very large increase in the complexity of the nervous system occurs with the appearance of the Platyhelminthes or flatworms. Ganglia in a flatworm are fewer

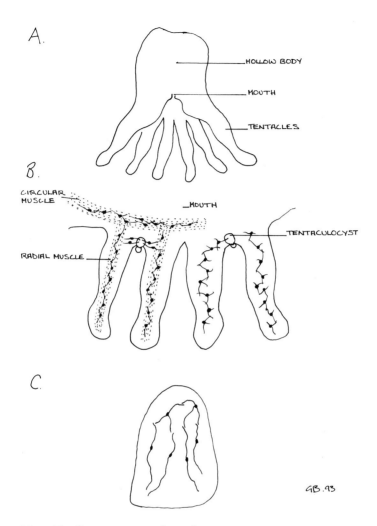

A.
HOLLOW BODY
MOUTH
TENTACLES

B.
CIRCULAR MUSCLE
MOUTH
TENTACULOCYST
RADIAL MUSCLE

C.
GB.93

Figure 4.2   The first appearance of ganglia.

in number and concentrated at one end of the animal, which may be distinguished as the head because primitive eyes and a mouth are found there. Shown in Figure 4.3 is the dorsal nerve plexus of *Notoplana atomata* (*Polycladida*) converging on the head ganglion, which actually consists of two interconnected ganglia. The labels refer to the gpl (genital nerve plexus), hdn (posterior dorsal nerve) and tau (tentacle eyes).

The grouping of neurons into a head ganglion that provides an integrating center for the nervous system reaches its most complex level of evolution in the

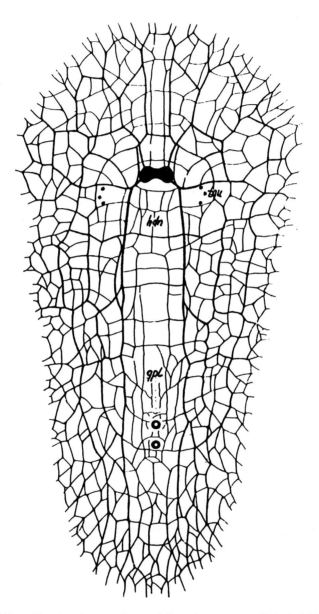

Figure 4.3   The dorsal nerve plexus of *Notoplana atomata (Polycladida)* converging on the head ganglion.

brain of *Homo sapiens*, which receives sensory information from nerve receptors in different areas of the skin called dermatomes as well as from the distance receptors such as the eyes, ears and nose.

Shown in Figure 4.4 are the dermatomes that bring information to the brain concerning such sensations as touch, pressure, temperature and pain. The individual areas are labelled C2 to C5 (cervical spinal cord levels 1 to 5), T1 to T12 (thoracic spinal cord levels 1 to 12), L1 to L5 (lumbar spinal cord levels 1 to 5) and S1 to S4 (sacral spinal cord levels 1 to 4). Nerves enter the spinal cord at each of these levels C2 to S4.

4.2 Humphrey's concept of the evolution of the sentient loop

Nicholas Humphrey[1], at one time the Director of the Unit of Animal Behaviour at Cambridge University, has put forward the following theory on the evolution of consciousness.

In the most primitive animals a nervous system consists of sensory neurons bringing in information (concerning, for example, touch and pressure) to an integrating center consisting of a large number of interconnected neurons constituting a head ganglion. The head ganglion then issues a motor command to contract an appropriate muscle given the type of sensory information it has received (Figure 4.5A). The next level of sophistication (Figure 4.5B) was reached with the appearance of collateral nerve branches (collaterals) emanating from the outgoing motor nerves, and ending in relation to the incoming sensory nerves; in this way the motor command was able for the first time to modify the sensory input to the head ganglion (an example of this process is described in relation to Figure 4.6). These collaterals then became modified in two important ways, one of which is shown in Figure 4.5C. On issuing a motor command the collateral is able to induce a sensory experience, independent of any input to the head ganglion, along the sensory nerves themselves (an example of this process is described in relation to Figure 4.7). The other important way in which the collaterals became modified (Figure 4.5D) is that they could be used to alter incoming sensory signals independently of any motor signals at all (Figures 4.8B and 4.11 give examples of this process). Figure 4.5E shows the final level of sophistication. This involves the appearance in evolution of the 'sentient loop', in which the collateral acts on its own without any motor command being issued or sensory information about the environment being received. The head ganglion, or brain, can in this way generate its own sensory experiences and maintain them at will. It is this ability to use the sentient loop that is the highest form of consciousness. (For an example of the brain generating activity in a voluntary way, without motor or sensory activity, see the discussion relating to Figure 4.11.)

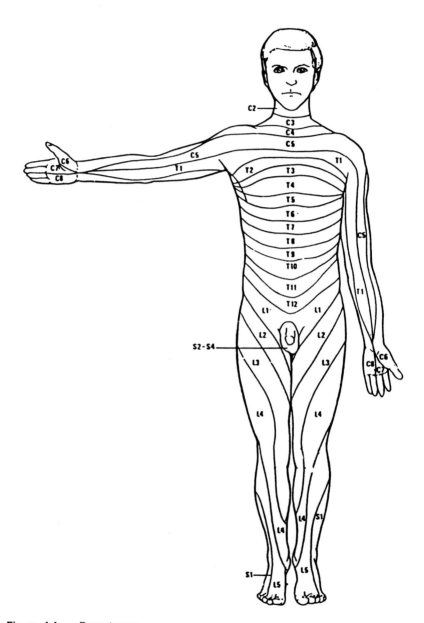

Figure 4.4   Dermatomes.

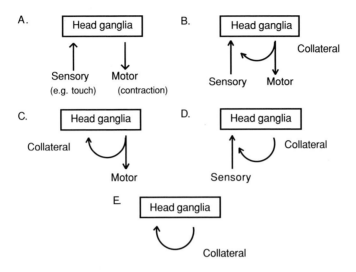

Figure 4.5   Evolution of the 'sentient loop' according to Humphrey.

4.3 The modification of sensory signals before they reach consciousness

What are collateral effects and can they operate in such a way as to modify sensory signals? One of the simplest motor acts that engages the brain and the spinal cord is shown in Figure 4.6 and illustrates how collateral effects can modify sensory signals before they reach consciousness.

A sensory stimulus, such as that arising from sensory spindle receptors in muscle cells (which are specialized nerve endings in particular muscle cells that sense the extent of stretch or speed of stretching of the muscle cell) is relayed through the sensory neurons just outside the spinal cord to the nerve cells just inside the cord within an area called the *substantia gelatinosa*. These signals are then sent to the group of neurons called the *nucleus gracilis* and *nucleus cuneatus*. From there the signals are sent to the thalamus in the brain, which is the receiving area for nearly all the sensory input to the overlying mantle of the brain, the cortex. Finally, the thalamus projects the information to that part of the cortex which is called the somatosensory area and is concerned with the analysis of information derived from sensory receptors in the limbs.

The kind of information in the signals processed by the somatosensory cortex may require that the muscles that gave rise to the sensory input in the first

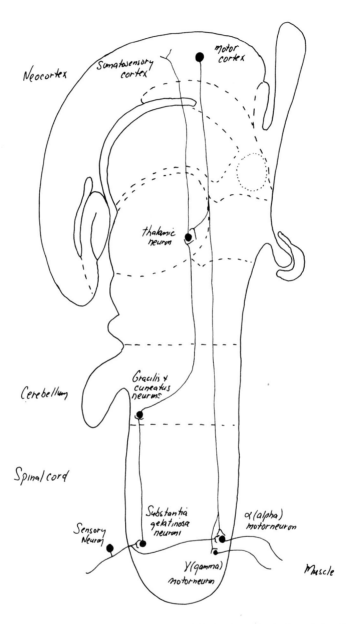

Figure 4.6   Diagrammatic representation of the brain and spinal cord showing the possible levels of corollary discharge by which motor output from the cortex acts on incoming kinesthetic signals arising from the sensory neurons.

place be contracted. In this case, neurons in the part of the cortex concerned with contracting muscles, namely the motor cortex, project a signal to the appropriate motor neurons in the spinal cord with synaptic connections to these muscles. These are of two different kinds: alpha motor neurons that have synaptic attachments to cells in the muscles in question that produce the force, and gamma motor neurons that have synaptic connections with cells that contract in the sensory receptor apparatus itself. Contraction of this latter type of cell changes the characteristics of the sensory receptors so that they rapidly send signals to the sensory neurons outside the spinal cord and from there to the alpha motor neurons, leading to the contraction of the bulk of the muscle cells. The sensory receptors also send signals to the somatosensory cortex via the thalamus along the pathway already described.

Two kinds of information are then sent via the sensory neurons to the brain. One of these relates to the signal arising from the sensory receptors in the muscle concerned with the position, tension and movement of the muscle, known collectively as kinesthesia. The other relates to the signal arising from the receptors as a consequence of their being contracted by the gamma motor neurons. This latter signal is removed (that is, gated out) before it reaches consciousness by a collateral from the motor pathway (the collateral branch to the thalamus in Figure 4.6). The sensory signal concerned with the state of kinesthesia of the muscle is not gated out, but is allowed to reach consciousness. The level in the brain or spinal cord at which this gating procedure is carried out is not known; the collateral branch is depicted at the level of the thalamus in Figure 4.6 simply for the sake of definiteness. Sensory signals can therefore be modified before they reach consciousness as a consequence of a collateral effect from the motor pathway.

4.4 Collateral effects can be used to generate sensations

Collateral effects (also known as corollary discharges) from the motor pathway can also be used to generate a sensation independent of any incoming sensory signals. They can generate sensations of muscular force or heaviness, although they cannot generate sensations of movement.

Figure 4.7A shows how the sensation of the heaviness of an object held in the hand is generated by a collateral effect. The pathway from the motor cortex to the alpha motor neurons is shown to give off a collateral branch at the level of the basal ganglia (depicted at this particular level for definiteness only). It is hypothesized that the associated corollary discharge generates a sensation in the cortex of the degree of heaviness of an object by firing impulses in proportion to those that are being propagated down the motor pathway. The neurons in the motor cortex that project to the motor neurons in the spinal cord to contract muscles involved in

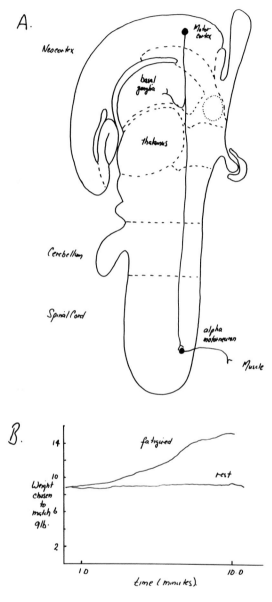

Figure 4.7   Diagrammatic representation of the brain and spinal cord illustrating a possible output from the motor cortex responsible for the perception of heaviness of a held object.[2]

lifting an object are called Betz cells. When these are activated a collateral signal is sent which gives rise to the perception of heaviness of the object being lifted (the collateral effect). It follows that when a muscle is weakened by fatigue, such as when holding a heavy suitcase, a greater number of impulses are required by the non-fatigued component of the muscle in order for the muscle to continue lifting the suitcase, and so more impulses are being propagated down the motor pathway. The collateral branch then receives a greater number of impulses and so a greater sensation of heaviness is experienced.

Stroke victims may have to send a larger than normal discharge down the remaining functional Betz cells to achieve the aim of, for instance, lifting their arms; therefore they experience their arms as an enormous burden.

An experimental example of this is given in Figure 4.7B. The graph shows the results of an experiment in which the subject had to support a nine pound weight with one arm (the experimental arm) while being asked at intervals to choose what they thought were equal weights to be supported in the same way by the other arm (the control arm). When the experimental arm was allowed to rest between the trials, the subject chose weights with the control arm close to the nine pound weight held by the experimental arm (see 'rest curve'). If, however, the experimental arm had to support the nine pound weight continuously, then the subject chose weights with the control arm that were successively greater (see 'fatigued' curve) than the nine pound weight, indicating the increased sense of heaviness. This arises from the increase in collateral effect with time associated with the muscles having to receive a greater number of impulses to support the weight continuously. The perception of heaviness does not arise from sensory signals in the muscle being relayed back to the brain.

Humphrey[1] is correct then in his suggestion that collateral effects can modify both the kinds of sensations that enter consciousness as well as generate sensations that do not arise from the workings of our sensory receptors. Figure 4.8 summarizes the situation. Humphrey speculates that early during evolution nerve pathways were layed down that allowed an animal to respond to, say, a noxious stimulus to the skin by 'wriggling' away. Figure 4.8A illustrates the simplest pathway involving the brain in this kind of motor pathway response. Primary sensory neurons relay information concerning kinesthesia, temperature and touch via the spinothalamic tract to the reticular formation of the hindbrain. Here a reflex act is initiated by exciting motor neurons to contract muscles in relation to the sensory stimulus.

At a later stage of evolution, mechanisms were put in place that allowed the nervous system to remove sensory information, using projections from the brain to the sensory gateway to the cortex, namely the thalamus, as shown in Figures 4.8B and 4.8C. We have already seen how information gathered by primary sensory neurons concerned with muscle receptors can be gated out before it

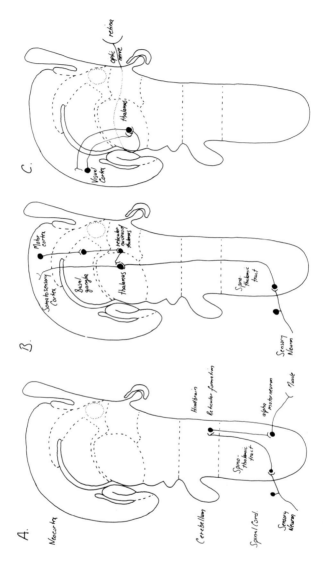

Figure 4.8 Diagrammatic representation of the brain and spinal cord showing different kinds of interactions between sensory pathways and motor pathways.

reaches the somatosensory cortex by means of a collateral feedback from the motor cortex at the level of the thalamus. Such a feedback could occur via the well-known pathway from the motor cortex to the basal ganglia and from there to the reticular nucleus (that lies just outside the thalamus) which then projects to the somatosensory cortex (Figure 4.8B). Sensory information that is gathered by the retina is also 'gated' as it passes through the thalamus on the way to the visual cortex, as shown in Figure 4.8C. The primary visual pathway is from the retina to the thalamus and from there to the visual cortex; neurons exist in the cortex that project back to the thalamus where they can gate the incoming visual information. There is then evidence for the modulation of both signals arising from the primary sensory neurons, and signals arising from the visual sensory neurons at the level of the thalamus. In this way the brain itself can determine the sensory information which reaches it and so modulate the perceptions of the world which it might allow to reach consciousness.

4.5 The consciousness of time

The example discussed in Section 4.4 of collateral effects generating the experience of heaviness when independent of any incoming sensory impulses shows that the modulatory effect of collaterals on the sensory information passing through the thalamus does not simply involve a gating operation; that is, the removal of information. Humphrey suggests that collaterals may also sustain impulse traffic in sensory nerves after the sensory perception has passed. The experience of sensations in the absence of continuing stimulation of sensory receptors (although these receptors are involved in the initiation of the experience) would result. This brings up a question concerning the time over which consciousness of a sensation occurs. The subjective nature of the experience of time in consciousness as compared with objective time, measured by a clock for instance, is well illustrated by the 'cutaneous rabbit' perceptual illusion.[3]

Figure 4.9A illustrates how a sequence of taps were delivered (1 to 10 on the left) on a subject's arm at one-tenth of a second intervals. The first five taps occur at the wrist , the next two on the forearm near the elbow, and the last three at the shoulder region (the subject was not allowed to observe these procedures). Surprisingly, the subject experienced the second tap as displaced from the wrist and the rest of the taps at equal distances along the length of the arm (at 1 to 10 on the right). It is in this sense that the brain interprets the taps as if an animal (rabbit) had run up the arm.

Five taps were then delivered in the sequence shown in Figure 4.9B; namely, only on the wrist (on the left 1 to 5). Even more surprisingly than in the first case, the taps were all experienced as confined to the wrist without any of them appearing to be spread out along the arm. Why then in the first experiment did the brain interpret the taps at the wrist as having been experienced as spread out along the arm whereas in the second case remaining confined at the wrist?

With reference to Figure 4.9, a plausible explanation why the first five taps in A were experienced as distributed along the arm whereas the five taps in B were confined to the wrist is that the taps are not perceived simultaneously with the events. In a certain window of time (one to two seconds) the brain determines

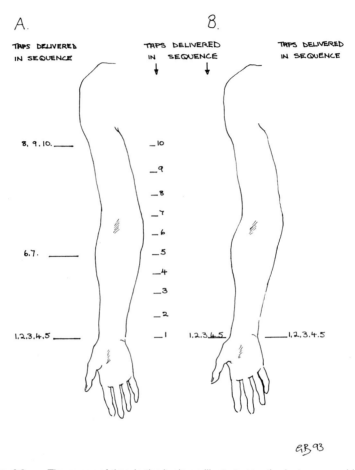

Figure 4.9 The sense of time in the brain as illustrated by the 'cutaneous rabbit' perceptual illusion.[3]

the most likely space–time story relating to the taps. In A the preliminary story that all five taps occur at the wrist is wiped out by the later arriving taps, so that the final story that enters consciousness is that the taps are spread out equally in a space-time sequence. In B the preliminary story that all five taps occur at the wrist is not wiped out by any later events, and so this interpretation enters consciousness. This illustrates how the brain uses the time available before behavior is acted out to arrive at the most reasonable story based on sensations (the taps) and past experience. This window in time could be delineated by the earliest time at which sensations enter the brain and the latest time at which the experiences might be used to modify behavior.

The actual time at which occurrences are first registered in the brain might not then be the same as the times allocated to them by consciousness. Another example of this is made with reference to Figure 4.10A, which shows the distribution of dermatomes for skin sensations as in Figure 4.4. The nerves leading from the dermatomes over the feet to the brain clearly involve a much longer pathway than do the nerves from the dermatomes over the neck to the brain. It might then be naively expected that simultaneous touches to the feet (at dermatome level L5) and the neck (at dermatome level C4), according to objective timing, would result in the experience of being touched on the neck entering consciousness before the experience of being touched on the feet. However, the actual time at which the occurrences are first registered in the brain is only part of the information that is used to allocate times to them entering consciousness. This also depends on the context in which this touching occurs as to whether one has the conscious experience of being touched in one place or the other within a certain window of time. The brain creates a story before it allocates a time to particular events. It does not simply take the actual time of arrival in the brain of impulses as if there was some finishing line in the brain which monitored the time at which the line was crossed by the impulses.

The hypothetical graph in Figure 4.10B illustrates that the time of experiencing being touched on various parts of the body (or on different dermatomes) need not coincide with the objective time of the sequence of touchings. The graph depicts the experienced times 1 to 5 for the objectively timed events of being touched on different sensory dermatomes in the spatio-temporal sequence S3 to C2 shown. The series of touches at one-fifth of a second intervals from S3 to C2 in the order shown may then be experienced as the temporal series 1 to 5 (that is, as a spatially continuous stroking from the buttocks to the head), depending on the story created by the brain, given the circumstances. The brain creates the most likely story, using the information that it receives from sensory receptors, the context in which this is gathered, and past experience, before allocating times to particular events.

Humphrey suggests that collaterals not only gate incoming sensory activity, for example at the level of the thalamus, but that they can also sustain that activity after the sensory receptors are no longer stimulated. This would then generate a sensation that is extended in time within consciousness. It gives rise to an important idea in Humphrey's scheme, namely that of the 'sustained sentient loop', in which the issuing of an outgoing command over a collateral can result in a sensation that is extended over time in consciousness by the sustained activity of the collateral.

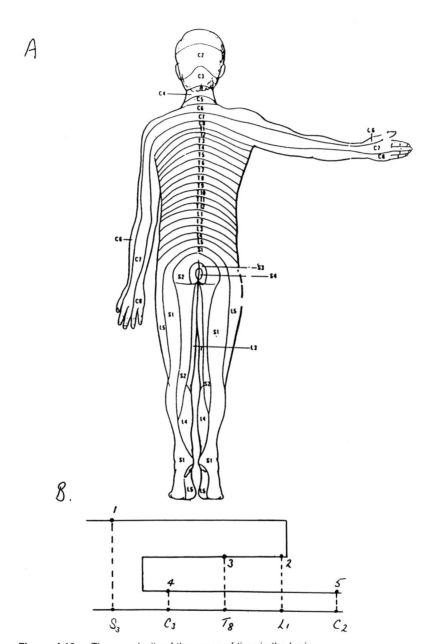

Figure 4.10   The complexity of the sense of time in the brain.

4.6 Brain activity in the absence of behavior

The brain possesses neurons which are active and which are not directly involved in either sensation or the issuing of a motor command. By monitoring the
consumption of analogues of glucose by neurons in different regions of the
brain with non-invasive techniques, it is possible to determine the areas of
neuronal excitability (for a more detailed description of this technique, see Chapter
5). Figure 4.11 shows how this technique has been used to determine the distribution of active neurons involved in the intention to perform a motor act. Active
neurons are found (as expected) in the motor cortex when a finger is flexed
against a spring; in addition, active neurons are found (again as expected) in the
somatosensory cortex, which is receiving kinesthetic information from the muscles being contracted (Figure 4.11A). However, if a more complex motor act is
executed, such as turning a key in a lock, then another set of active neurons is
brought into action in the area of the brain called the supplementary motor
cortex (Figure 4.11B). This area is always active when complex motor activity is
taking place. If the turning of a key in a lock is now simply rehearsed mentally
(by issuing commands associated with the task but not allowing them to be
carried out), then the supplementary motor cortex possesses active neurons as
before but the motor cortex and the somatosensory cortex do not (Figure 4.11C).
This is then an example of the motor system operating in the absence of any
motor output at all.

4.7 The evolution of the collateral effect

The central idea in Humphrey's scheme is that collaterals, originally associated
during evolution with the motor system, may give rise to a sustained sentient
loop without there being any motor act performed. According to Humphrey, to
feel a sensation in consciousness is to issue a command or outgoing signal;
sensation is then the making of the sensory response. We have seen that the
motor system itself, in the case of the supplementary motor cortex, may give rise
to activities that do not result in a motor action. The issuing of commands that
set up a sentient loop amounts to the experiencing of sensations over time; this
is a process that has become modified from the original collateral effects, which
simply acted on incoming sensory information.

Humphrey's ideas concerning the evolution of the sustained sentient loop,
and therefore consciousness, are summarized in Figure 4.12. At first there was a
simple nerve pathway consisting of a sensory input, which might be related to a
noxious stimulus to the skin, resulting in a motor output involving withdrawal
from the site of the stimulus. In Humphrey's terminology this amounts to a
'wriggle of rejection', which he traces back to simple animals like sponges. It is

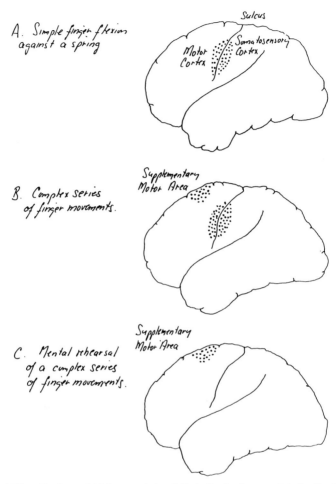

A. *Simple finger flexion against a spring*

B. *Complex series of finger movements.*

C. *Mental rehearsal of a complex series of finger movements.*

Figure 4.11   Regions of high neuronal activity in the brain associated with simple and complex motor (muscular) tasks and with the rehearsal of motor tasks without any muscular activity.

shown in Figure 4.12A as involving the brain, but it would be better represented in vertebrates by a reflex sensory nerve pathway that passes directly from the skin to motor neurons in the spinal cord and from there to the appropriate muscles, as in Figure 4.6. The next stage in the evolution of the sentient loop involves modification of the incoming sensory signal by a collateral effect (or corollary discharge) from the outgoing motor signal, as in Figure 4.12B; examples of this occur in the removal of components of the signals to do with the action of muscle receptors involved in gamma motor neuron activity by motor collaterals, as discussed in relation to Figure 4.6.

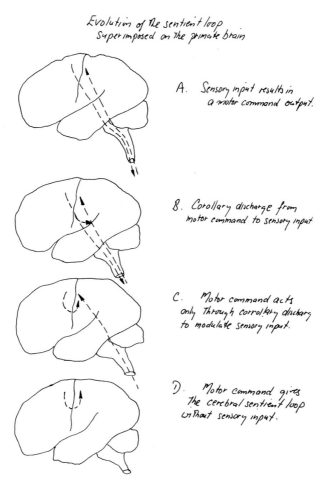

Evolution of the sentient loop
Superimposed on the primate brain

A. Sensory input results in
   a motor command output.

B. Corollary discharge from
   motor command to sensory input

C. Motor command acts
   only through corollary discharg
   to modulate sensory input.

D. Motor command gives
   the cerebral sentient loop
   without sensory input.

Figure 4.12 Evolution of the sentient loop, and therefore consciousness, as envisioned by Humphrey and superimposed on the primate brain.

With the further evolution of collateralization, the motor command could modify and sustain over time the information coming into the brain along a sensory pathway so as to sustain a sensory experience (as shown in Figure 4.12C). More explicitly, Humphrey suggests that the corollary discharge associated with a motor command in the context of a particular sensory experience may become modified so that the motor command is not executed and the corollary discharge is then used to sustain in subjective time the sensory experience. This gives the 'after glow' of a sensory stimulus; that is, the experience is maintained in subjective time even though it has passed in objective time. The projection from the motor cortex to the reticular nucleus of the thalamus provides just such a pathway for modifying and sustaining the sensory input arriving from primary sensory nerves (as discussed in relation to Figure 4.8B).

Finally, the stage is reached during evolution when collaterals are used to generate sensations, independent of any sensory input to the brain, as in Figure 4.12D. The cerebral sentient loop is now independent of the environment, and the experience of a sensation involves a positive act of issuing an appropriate outgoing signal from the brain. According to Humphrey, sensing is not a passive act but involves participating in the act of 'sentition'. That is, sensing involves the issuing of an appropriate outgoing signal from the brain; a signal which was, in its first evolutionary appearance, associated with the motor system only. It is this process which Humphrey claims to be consciousness. Since these commands can be issued without any trigger from the environment it is possible to have a rich 'stream of consciousness' that is generated from within the brain itself.

4.8 Consciousness as a consequence of the sentient loop

The present chapter has looked at collateral effects and feedback pathways that modify incoming sensory signals bringing information about the environment. The idea of sentition, whereby the nervous system issues a command that results in a sensory experience and therefore consciousness, is a novel one. According to this idea, consciousness first appears during evolution with the species that uses motor collaterals to generate or modify sensory inputs to the brain. If it can be shown that flatworms are able to modify the sensory input to their central head ganglia by means of motor collaterals, then this occurred with their first appearance. Any animal that can issue commands for altering or generating sensory activity, and can by this means make a sensory response, possesses consciousness according to this idea. This does have the great attraction of providing some basis for continuity in the emergence of consciousness, rather than just positing it as the special preserve of *Homo sapiens* or even of just the mammals, a problem that is returned to in Chapter 6. The deficiency in the idea of sentition is that it does not provide a framework that is sufficiently specific to suggest a research plan that allows testing the central hypothesis of the sustained sentient loop as the basis for consciousness. Although consciousness can only be examined by introspection, the non-invasive techniques for examining the neurophysiological concomitants of mental functioning, such as positron emission tomography, may help to clarify the issues, as is argued in Chapter 5. It will be interesting to see if, under suitable conditions, those areas of the brain involved, for example, in forms of cognition that do not involve language are also active in mammals other than the primates, a problem considered further in Chapter 6.

4.9 References

1. Humphrey, N. (1992) *A History of the Mind*. London: Chatto & Windus.
2. McCloskey, D. I., Ebling, P. and Goodwin, G.M. (1974) Estimation of weights and tensions and apparent involvement of a 'sense of effort'. *Experimental Neurology*, **42**, 220–232.
3. Geldard, F.A. and Sherrick, C.E. (1972) The cutaneous 'rabbit': a perceptual illusion. *Science*, **178**, 178–179.

4.10 Further reading

Blakemore, C. and Greenfield, S. (1987) *Mindwaves*. Oxford: Blackwells.
Ciba Foundation (1993) *Experimental and Theoretical Studies of Consciousness*. New York: Wiley.
Dennett, D.C. (1991) *Consciousness Explained*. London: Penguin Press.
Edelman, G. (1992) *Bright Air, Brilliant Fire*. London: Penguin Press.
Gregory, R. (1988) *The Oxford Companion to the Mind*. Oxford: Oxford.
Searle, J. (1992) *The Rediscovery of the Mind*. Cambridge: M.I.T. Press.
Scientific American (1992) 'Mind and Brain', Special Issue, **267(3)**.

CHAPTER 5   THE DISTORTION OF CONSCIOUSNESS

5.1 Introduction

The formation of a mature human brain involves genetic and environmental factors which together result in a neural machinery that operates in an holistic fashion in consciousness. This seems to occur by means of the synchronization of the output of what are known as 'distributed modules' throughout the cortex. Distributed modules are equivalent to neural circuits in the brain that subserve different functions, generally in parallel; functions such as those involved in the holistic process of the visual experience of a person walking past, carrying some roses. Here, the synchronization of the output from movement modules, object recognition modules, olfactory modules, and many others gives rise to the holistic visual experience.

But we do not know how these modules malfunction and give rise to the distortions of consciousness involved in such psychotic disorders as schizophrenia and bipolar manic depression. We do not know what has happened to the circuitry of the module to cause, for instance, the person no longer to be seen as moving, though they can still be seen and the roses they carry can still be smelt. The study of schizophrenia is of particular interest as this mental condition involves changes to consciousness that are at the very essence of what it is that makes a person an individual; for instance, to personality and personal experience. Such changes are far removed from mere sensory input. Any insights obtained as to the origins of the different kinds of schizophrenia offer not only the possibility of contributing to the alleviation of this condition, but also of coming to grips with the neural basis of consciousness itself.

Schizophrenia afflicts at least one in every hundred of the population throughout the world and is responsible for one in three admissions into mental hospitals. With between 20 and 40 million people suffering from this mental disturbance of consciousness, schizophrenia is the most serious unsolved disease in world society, a view supported by *Medical Research: A Midcentury Survey*, sponsored by the American Foundation.

To be schizophrenic, as defined by the *Diagnostic and Statistical Manual of Mental Disorders* of the American Psychiatric Association, is to experience symptoms that last for at least six months and include at least two of the following: delusions, hallucinations, disorganized speech, grossly disorganized or catatonic behavior, negative symptoms. Negative symptoms include restrictions in the range and intensity of emotional expression (affective flattening), in the fluency and productivity of thought and speech (alogia), and in the initiation of goal-directed behavior (avolition). The term 'negative' is used to contrast these aspects with the positive aspects which produce something, namely hallucinations. The kind of delusions that are most common are those associated with some kind of persecution, such as being spied on, ridiculed or tormented in some way, and often involve grandiose misconceptions. The most common type of schizophrenia is the paranoid type, in which the delusions are accompanied by auditory hallucinations and, to a much lesser degree, visual hallucinations. Most importantly, such schizophrenics do not have disorganized speech, and do not possess disorganized or catatonic behavior, nor do they exhibit flat or clearly inappropriate feeling or emotion. Neuroscientists can therefore investigate paranoid schizophrenia with hallucinations in a single sensory modality in a way that is not further complicated by these other conditions. Because every individual's own stream of consciousness is so intensely personal, the condition of auditory hallucinations in which two or more voices are conversing with one another, or maintaining a running commentary on the thoughts within consciousness, is of particular interest. This chapter considers the possibility of gaining insights into the mechanisms by which this comes about; that is, the mechanisms which generate voices distinct from those ordinarily known as 'our own thoughts'.

The inquiry as to how the brain gives rise to hallucinations has fascinated humanity since before Aristotle. The fifteenth century Aristotlian concept of the brain comprised of modules known as 'cerebral cells' in which different mental functions were believed to be carried out. There was, for instance, a module for memory as well as one for fantasy. The brain was not considered to be the sole repository of mental functions as there were also modules in the heart for the sensations arising, for example, from touch and taste.

Nowadays the study of the neural basis of auditory or visual hallucinations (and neuroscience in general) proceeds at various levels. Neurobiologists investigate at the level of neurons. They probe the function of the connecting synapses and the molecules of transmitter substances that are released at these synapses which mediate the transfer of information at the synapse. Thus

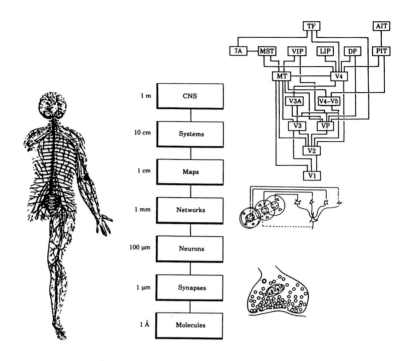

Figure 5.1   Different levels for the consideration of brain function.

neurobiologists are concerned with objects of study with dimensions on the nanometers to micrometers scale, as indicated by the bottom three rectangular boxes in Figure 5.1. A synapse is shown to the right of these boxes.

Neurophysiologists investigate the problem at the systems level, such as trying to understand how the auditory system or the visual system operates. They study the patterns of connectivity of the nervous system within the separate systems that compose it. There are, for example, over thirty distinct areas in the brain that generate consciousness of the visual world. One such network of connections is that between lateral geniculate neurons in the thalamus and the primary visual cortex (V1) in the occipital cortex; this network is illustrated in a drawing by Hubel and Wiesel, the discoverers of these connections, on the right of the 'Networks' box. The interconnections of networks (such as those connecting V1 with other sets of networks carrying out different roles in the process of analyzing the visual world) are indicated in the drawing to the right of the 'Systems' box (with the different networks labelled V1, V2, et cetera). Studies at this level are of neural networks in the dimensional range of millimeters to centimeters.

A final level of investigation is indicated by the rectangular box labelled CNS for central nervous system. The analysis of the intact CNS, especially that of humans, is primarily the province of those carrying out non-invasive studies of brain function, such as people engaged in psychophysics and brain imaging.

The problem of the origins of hallucinations is studied at the systems level using techniques for imaging different intensities of activity in different areas of the brain in a non-invasive way. The other main approach involves identification of those molecules in the brain which are likely to be inappropriately distributed and so give rise to the hallucinations which accompany conditions such as schizophrenia. Both of these approaches will be elaborated on in this chapter.

5.2 Imaging brain functions related to consciousness

Neurons that are undergoing enhanced levels of electrical activity consume more glucose. Analogues of glucose, such as 2-deoxyglucose, are taken up by neurons and phosphorylated in the same way as glucose. The product, deoxyglucose 6-phosphate, accumulates in neurons so that a measure of the electrical activity of a particular neuron is the extent to which it has accumulated deoxyglucose 6-phosphate. This can be ascertained by first attaching the positron-emitting isotope fluorine-18 (^{18}F) to deoxyglucose to give ^{18}F-labelled deoxyglucose. Subsequent introduction of this into the bloodstream serving the brain ensures that the activity of brain neurons can be determined if their relative accumulation of ^{18}F deoxyglucose 6-phosphate can be measured. This is done by detecting the levels of positron emissions from the labelled neurons in different parts of the brain. The detection is achieved as a consequence of the emitted positron colliding with an electron, thus giving rise to two gamma rays that propagate in opposite directions from the ^{18}F deoxyglucose 6-phosphate. Detection of these two simultaneously emitted gamma rays by detectors placed around the skull allows for the site of the radioactive deoxyglucose to be determined to within a few millimeters; hence the rate of brain metabolism and therefore electrical activity at the site can be calculated. This technique is called positron emission tomography (PET).

Another non-invasive technique for imaging the brain is provided by functional magnetic resonance imaging (fMRI). This relies on the process whereby the imposition of a magnetic field across the brain gives rise to the alignment of the direction of spin of the nuclei of such atoms as hydrogen, which when perturbed by a brief pulse of radio waves tip their direction of spin. On removal of the pulse, the spin directions realign with the magnetic field and in doing so release radio waves. Because the frequency of these radio waves is different for different atomic species, as well as for a particular type of atomic nuclei in different environments, it is possible to use these radio wave emissions to differentiate between, say, the neuron cell bodies and their myelinated axons, or between the water in the cerebrospinal fluid and that in the brain. Changes in the amount of oxygen carried in the blood alters that fluid's magnetic properties, which can be detected by fMRI. Because the amount of oxygen in the blood in the area of neuron activity changes, and this can be detected with fMRI, neuronal activity in a particular part of the brain can be localized. This measurement with fMRI has an accuracy of a millimeter or less. It therefore provides another non-invasive method for measuring the relative activity of neurons in a particular part of the brain, as these changes reflect local electrical activity of neurons due to their changed oxygen requirements.

The power of these non-invasive imaging techniques to locate those parts of the brain that are engaged in producing some experience in consciousness is vividly illustrated by consideration of the phenomenon called the waterfall effect. This name derives from the fact that if a subject looks for some time at water streaming downward in a waterfall and then turns away and looks, for example, at the opposite bank of the river, then the trees on the bank will appear to be momentarily moving in the opposite direction to the flow of the water in the waterfall. Thus the stationary objects (the trees) undergo an illusory motion. The time course of this illusory experience, in which subjects think they see motion whereas in reality the stimulus is stationary, can be described quantitatively as shown in Figure 5.2.

The cortical area in the visual pathway called MT is known to respond better to moving than to stationary stimuli. Psychophysical experiments[1] in conjunction with brain imaging can be used to show that this part of the brain is involved in producing a visual illusion in consciousness. Figure 5.2Aa shows the visual stimulus which consists of a series of concentric rings, either expanding or stationary. Figure 5.2Ab indicates the part of the brain (MT) that is excited by the moving rings as determined by fMRI (note that the brain is shown in normal and inflated format). The same increase in activity in MT is observed when stationary concentric rings are viewed immediately after observing the rings expanding, but not when observing just stationary rings without prior exposure to moving rings, or after prior exposure to rings moving in one direction and then reversing and moving in the opposite direction. It is known that the experience in consciousness is of a perceptual motion after-effect in the first case but not in the latter two cases; that is, the stationary rings appear to move in the opposite direction to that used in the conditioning period (the period of expanding or contracting stimuli) for some time after the conditioning period.

Figure 5.2B depicts the changes in the fMRI signal during real and illusory visual motion. The strength of the signal in area MT of the brain is shown for the case of the moving concentric rings either continuously expanding (Exp),

continuously contracting (Con), reversing direction (expanding then contracting, Exp/Con) or stationary (Stat). Following the periods of continuous
unidirectional local motion (the expanding or contracting stimuli), a visual motion after-effect is seen by the subject in the physically stationary rings. Following
the period of reversing the direction no motion after-effect was reported by the
subject. The fMRI response during the period of expansion of the rings, contraction of the rings, or reverse direction expansion/contraction is about the
same (between 3.0 and 3.5). The important point to notice is that the fMRI
response immediately after the single-direction stimulus (that is, when the motion after-effect was being experienced) remained high for a considerable period
of about twenty seconds after the stimulus offset, much longer than that of the
case when there is no after-effect, such as the case of reverse expansion/con-
traction in the stimulus for which the fMRI signal returns to the no stimulus
base-line in about five seconds. Thus MT remains active during the period
when there is no stimulus but the subject experiences the visual after-effect.

The final graph (Figure 5.2C) shows the quantitative relationship between
the visual after-effect and activity in MT as measured by fMRI (compare with
Figure 5.2B). The line gives the fMRI amplitudes during and after single direction expansion of the concentric rings minus the amplitudes during and after
reversing-direction conditions. During the first forty seconds there was no difference between the activation produced by single direction versus reverse
direction stimuli (see Figure 5.2B). During the next forty seconds, the fMRI
results show that as the subjects observed stationary stimuli the fMRI was
elevated for about twenty seconds following the single direction stimulus. These
results may be compared with the psychophysical data which is indicated by
the open squares. These measured the period as reported by the subject of the
time course of the visual illusion which was experienced whilst observing the
stationary rings in a related series of psychophysical experiments. There was
very good agreement between the time course of the decline in the visual illusion given by the open squares and that of the activity in MT as indicated by
the fMRI. The results show that illusions have a neural representation in the
brain, and in the case of the waterfall effect this has been identified with MT.

5.3 Problems with interpreting brain images related to consciousness

It might be anticipated that these techniques of non-invasive brain imaging
could be used in a relatively straightforward way to determine those areas of the
brain that are producing the distortions of consciousness that occur in the
auditory hallucinations of schizophrenics. That this is not the case is best illustrated by consideration of the following fundamental problem. When one
perceives an object, for example the Sydney Opera House, then the image of the

A

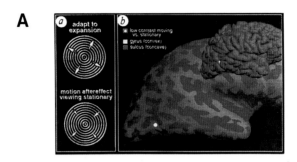

B

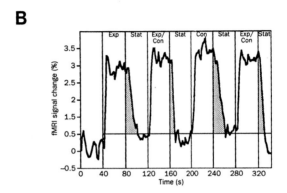

C

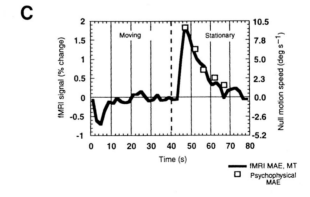

Figure 5.2 Psychophysical experiments in conjunction with brain imaging used to show the part of the brain that experiences a visual illusion in consciousness.[1]

Opera House on the retina must be reconstructed in the cortex. The requirement of activity in V1 is firmly established and has been described in Chapter 1. This is the well researched process involving the forward projection from the retina to the thalamus and from there to V1 in the occipital cortex and beyond, as shown by the forward projecting arrow in Figure 5.3. What happens when the eyes are now closed and the structure of the Opera House is imagined instead of actually being perceived? Does V1, involved in the reconstruction of the building from the information provided from the retina, now again reconstruct the Opera House? If this is the case, then it may occur as a consequence of information reaching V1 from higher visual centers such as those in the temporal lobes. This would require a backward or downward projection to V1 as indicated by the arrow in Figure 5.3. Such a problem would appear to be resolved easily using PET or fMRI. Just ask subjects to view the Opera House and measure the regions of enhanced neural activity in their brains; this will involve V1. Then ask them to close their eyes and imagine the Opera House, at the same time measuring the neural activity in this way again to see if it is high in V1. The results of this procedure turn out to be much more complex to interpret than anticipated.

The use of PET has lead to general agreement that both the parieto-occipital as well as the temporo-occipital visual association areas of the cortex are involved in imagining a visual scene such as the Opera House. This even holds for more complex imaginations, such as those involved when imagining the action of walking out of the front door and then walking alternatively to the left and to the right upon reaching successive corners. However, whether V1, which

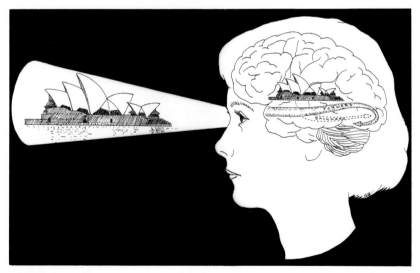

Figure 5.3   Projection to the primary visual cortex.

receives a direct input from the thalamus, is activated is unclear[2]. The main protagonists are leaders in the imaging field, Roland and Kosslyn. Roland and his colleagues argue that V1 is not involved in the reconstruction of the image in consciousness. In imagined scenes in which a subject performs a walk of the kind just described, the parietal and temporal visual association areas are clearly excited according to changes in regional cerebral blood flow, but not V1[3]. On the other hand, Kosslyn and his colleagues contend that V1 is involved in imagery. Their subjects were asked to close their eyes and visualize an upper case letter of either a larger or smaller size, then hold this image and report on its character-istics, for example whether or not it possesses curved lines. PET reveals a very interesting result; the activation of the posterior portion of the medial occipital lobes is evident for the smaller letters with the activation towards the anterior portion of the medial occipital lobes occurring for the larger images. Since V1 is mapped topographically from the retina, with the fovea of the retina that would be used to perceive the smaller letters when the eyes are open represented at the posterior portion, the results make sense: the posterior portion of V1 is activated when there is an image of a small letter but the anterior portion is activated when there is an image of a large letter[4]. Furthermore, Kosslyn and his colleagues found that the activity in V1 is directly related to the size of the image, as would be expected given that this early area of visual cortical processing is topographically organized.

To what extent can the claims of the medical profession that a particular area of the brain is involved in schizophrenia, as determined by PET, be relied upon? The chances that correct identification can be made of the areas of the brain in-volved in, say, the visual hallucinations of schizophrenics seem a bit slim, given that the fundamental question as to whether V1 is involved in the reconstruction of visual images is not agreed upon by experts. Kosslyn points up the probable basis for his disagreement with Roland as residing in what are called the baseline conditions in setting up PET or fMRI studies. These turn out to be very complex. For example, in Roland's research the activation of the brain in the imagery test is compared with the activation present when the subject is not performing a re-quested imagery task, but simply lies motionless with eyes closed. However, this does not guarantee that the subject is not imagining something in visual con-sciousness; that is, just day-dreaming. As the detection of the increased activity in the imaging task requires the subtraction of this background activity, it is easy to see that activation of V1 during the day-dreaming episode could remove the indications of activity in this area of the brain during the imaging task. Kosslyn attempts to get around this problem by preserving the nature of the imaging task in both the baseline condition as well as during the imaging. For example, subjects are asked to close their eyes and imagine a letter of the alphabet, with the relative activity in the brain at this time taken as the baseline condition. Manipulation of this image in consciousness or reporting on detailed properties of the letter during the test conditions further enhances the activity in the brain. Consequent subtraction of the brain activity in the baseline from the test conditions consistently shows an enhanced activity in V1.

The discrepancy between the approaches of Roland and Kosslyn are further brought out in recent experiments[5]. In a resting baseline condition, subjects were asked to close their eyes (after being blindfolded with a dark cloth) and 'have it black in front of your mind's eye', a state that probably includes the generation of mental images. Subjects were also given an auditory stimulus to which they passively listened. Finally, in the imagery condition, the auditory stimulus was used to trigger the requirement that the subject produce an image in consciousness of a telephone. The regional blood flow in the brain was then determined for each of these three conditions using PET. The results were quite clear that subtraction of the activity measured during the so called resting baseline condition from that when the subject was producing an image did not reveal an increase in activity in V1. On the other hand, subtraction of the activity during the passive listening period from that of the imagery period did[5]. The experiments indicate that when the imagery condition is compared to that of the resting baseline, in which the brain activation is uncontrolled, imagery activation is obscured because of activation in V1 during the baseline condition. However, when the imagery condition was compared to a baseline in which the same auditory cue was presented, giving a controlled baseline, activity in V1 was evident. This example clearly demonstrates how difficult it is to set up the appropriate baseline conditions for analyzing the results of PET or of fMRI.

Even given these results of Kosslyn and his colleagues, it can be argued that V1 is not required for the task of generating images. For instance, some patients with cortical lesions are unable to identify real objects or the drawings of scenes by sight, but still have the capacity for vivid visual imagery. This implies that the visual processing involved in perception and in imagery are quite different[3]. Indeed, Crick and Koch[6] suggest that as all planning of a visual task takes place in the frontal cortex, then the contents of visual awareness in consciousness must be made available to the frontal lobes in order for this planning to proceed. Visual areas V4 and V5 possess such connections to the frontal cortex whereas V1 does not. (Note that V5 is commonly referred to as MT, especially in humans, and the distinction between V5 and MT is ambiguous.) It is therefore unlikely that visual consciousness resides in this area. The extent to which activity in V1 is a necessary concomitant of visual consciousness, either of a percept from a visual scene or of an image generated at will, is still unclear. What is clear is that great care must be taken in interpreting the results of the PET and fMRI techniques, to such an extent that it may be argued that the intellectual challenge in using such techniques really comes in the design of the appropriate baseline conditions to be used. It suggests that a healthy degree of scepticism should be exercised in taking at face value the various claims that are made as to the neural origins of such mental disturbances as those that arise in schizophrenia.

5.4 Brain images during the distortions of consciousness that accompany schizophrenia

Can visual hallucinations that occur during a schizophrenic episode illuminate the problem of whether activation of V1 occurs in visual imagery? Hallucinations in schizophrenia are predominantly auditory, although on rare occasions they may also be visual. The determination of enhanced activity in the brain of a patient undergoing visual and auditory hallucinations has been ascertained using PET.

The PET scans in Figure 5.4 (showing the surface projections with significantly increased activity during visual and auditory verbal hallucinations of the schizophrenic subject) indicate that the areas of activation are in the visual association cortex (specialized for higher order visual perception) and the auditory-linguistic association cortex (specialized for speech perception). Given the caveats just mentioned concerning the establishment of appropriate baseline conditions, it will be noticed that there is no activation of V1 (in the occipital cortex) in this patient. The internally generated visual images, which in this case took the form of moving colored scenes with rolling disembodied heads that spoke and gave instructions, involve a different pathway to that used when images are generated in a voluntary way. The study of this schizophrenic patient supports the view that activity in V1 is not necessary for visual experience[7].

Verbal hallucinations are the most common abnormality of consciousness in the paranoid schizophrenic brain. Figure 5.5 shows activity in the brain of a paranoid schizophrenic patient during a verbal hallucination as determined by PET. The hallucination took the form of voices talking to, or about, the subject. The various axial sections (at −16 mm, −8 mm, et cetera) show that there is increased activity in the thalamus, putamen, parahippocampus, hippocampus and anterior cingulate during the hallucination[7].

The visual and auditory association cortices are concerned with the internal production of visual and auditory phenomena. The cingulate cortex has major connections with the visual and auditory association cortices, allowing communication to occur between these brain regions. The primary involvement of the thalamus is of particular interest. Why should this structure, which seems overwhelmingly involved in the projection of sensory information to the cortex from the peripheral sensory organs, be in any way involved in verbal hallucinations? It may be that the thalamus, which receives massive nerve projections from the cortex as well as delivering its sensory information to the cortex from the periphery,

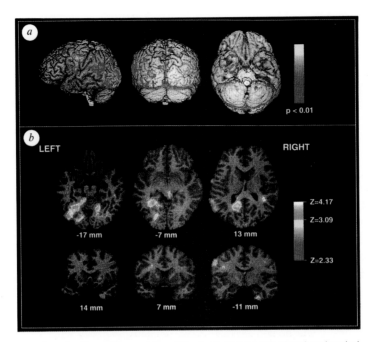

Figure 5.4 Localization of the regions of the brain involved in visual and verbal hallucinations in a schizophrenic.[7]

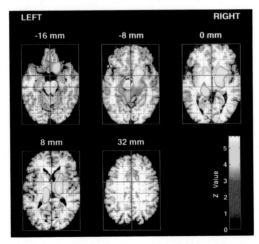

Figure 5.5 Localization of the regions of the brain involved in verbal hallucinations of a schizophrenic.[7]

is involved in the structuring of an internal representation of reality. This may be the case whether that representation is triggered by external events, or generated internally. In either case, the thalamus becomes vital for what is commonly called a 'grip on reality'. According to this idea, lesions here would lead to distortions in that reality through the generation of hallucinations.

A recent comprehensive study supports this view. The activity in an average schizophrenic brain, determined with fMRI, was reconstructed from a very large number of patients and compared with that of the activity in a large number of normal brains. Following subtraction of the two averages the area of difference was shown to be in the thalamus[8].

5.5 The dopaminergic hypothesis for the distortions of consciousness that accompany schizophrenia

Figure 5.6 shows the distribution of synapses formed by nerve terminals that release the transmitter dopamine, a catecholamine. This is of great interest because neuroleptic drugs, like haloperidol, used to alleviate the positive aspects of schizophrenia (such as auditory verbal hallucinations) block transmission at dopaminergic synapses.

A diagram of the three-dimensional relationship between the parts of the brain that contain dopaminergic neurons and their terminals is given in Figures 5.6A, B and C. The ventral tegmentum and the substantia nigra in the midbrain contain the cell bodies of these neurons. Those in the ventral tegmentum make a dense projection to the frontal lobes of the cortex as well as to the cingulate cortex (not depicted here as it is a deep structure best shown in 5.6B and C) and to the hippocampus (another deep structure also shown in 5.6B and C). The neurons in the substantia nigra project exclusively to the basal ganglia where they terminate in the different sets of neurons that compose this structure, namely the caudate nucleus, the putamen (P), the globus pallidus (GP) and the lateral medial (Lm) neurons. CC indicates the corpus callosum, T the thalamus and Sn the subthalamic nucleus.

Figure 5.6B gives a midline view of the distribution of dopaminergic nerve terminals from the ventral tegmentum. Two main nerve pathways are shown, one of which possesses terminals in the frontal lobes (FL) as well as to a lesser extent over the rest of the cortex. The other pathway possesses terminals in the cingulate cortex (C), above the corpus callosum group of axons that join the two hemispheres; this pathway extends around the cingulum to eventually reach the hippocampus (H). The amygdala is indicated by A; T indicates the thalamus.

Figure 5.6C is similar to Figure 5.6B but shows the projection from the substantia nigra to the basal ganglia; CC indicates the corpus callosum and Ce the cerebellum.

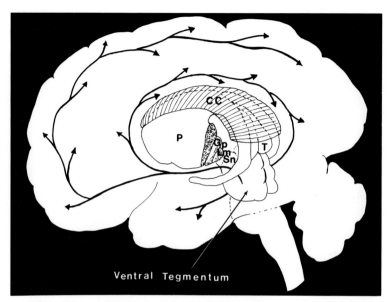

Ventral Tegmentum

Figure 5.6A Three-dimensional relationship between the parts of the brain that contain dopaminergic neurons and their terminals.

PET scanning can be used to examine the regulatory role of dopamine on cortical function. There is an impaired cognitive activation of the cingulate cortex in schizophrenics performing a verbal task involving word repetition. PET scanning shows that an agonist for dopamine receptors, such as apormophine, substantially improves the function of the cingulate cortex. The results suggest that in schizophrenic patients there is a failure of a normal dopaminergic modulation of neurons in the cingulate cortex. Because of the widespread connections of the cingulate cortex, this could have repercussions for the integration of activity for relatively remote regions of the cortex that are connected via the cingulate cortex[9].

The definition of schizophrenia, as already noted, is that a patient possesses two or more of the following symptoms: delusions, hallucinations, incoherent speech, disorganized or catatonic behavior, and negative symptoms (which include affective flattening, avolition and alogia). Planning and volition involve neural activity in the frontal cortex; if such activity fails, alogia and avolition can result. Because the cingulate cortex connects the areas of the associational cortex involved in the higher processes of verbal and visual experiences in consciousness, failure of this pathway can give rise to auditory and visual hallucinations. The hippocampus and amygdala are responsible for the laying down of declarative or specific memories and emotional memories, respectively. Thus failure of normal activity in, say, the amygdala can give rise to the flattening

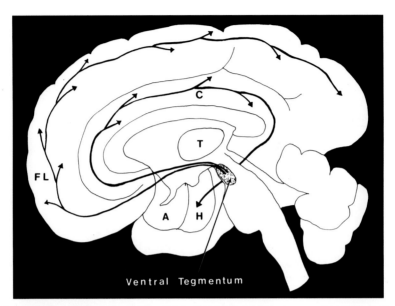

Figure 5.6B   Midline view of the distribution of dopaminergic nerve terminals from the ventral tegmentum.

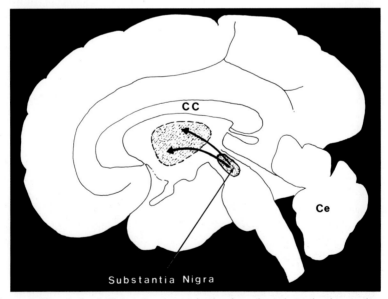

Figure 5.6C   As for 5.7B but showing projection from the substantia nigra to the basal ganglia.

of the emotions. Is there a common agent that could be malfunctioning in each of these brain areas, so giving rise to the schizophrenic condition? The answer is 'yes', for (as discussed in relation to Figure 5.6) each of them receives an extensive projection of axons from the ventral tegmentum of the midbrain; these release the transmitter dopamine. There is a very dense projection of these dopaminergic axons to the frontal cortex, to the cingulate cortex, and through the cingulate cortex to the hippocampus and the amygdala. Thus a malfunction in this ventral tegmentum projection can have severe repercussions for the workings of a range of activities associated with consciousness. These repercussions can give rise to the schizophrenic mind.

The greatest contribution to the alleviation of suffering associated with schizophrenia was the discovery by Laborit and his colleagues in Paris in the early 1950s of the actions of chlorpromazine. This agent brought several of the schizophrenic symptoms already discussed under control, so producing the most significant revolution in the treatment of the mentally ill of all time. It can be truly said that no progress had been made in the medical treatment of such patients since medieval times up until the year 1952. To see the suffering of a paranoid schizophrenic, subjected to the horrific visual hallucinations of disembodied talking heads criticising and admonishing, accompanied by verbal hallucinations consisting of demands that sufferers perform acts of violence, very often against themselves, is a distressing experience. The action of chlorpromazine quieting the hallucinations and giving respite from the condition appears like a miracle. Here then is an agent that acts on the contents of consciousness to specifically remove hallucinations. It allows the patient to distinguish a reality that enables interaction with the world in a recognizably social way.

Chlorpromazine is so central to the analysis of the mechanisms that are at fault in the schizophrenic brain that a brief history of its discovery needs to be made. In 1944, Paul Charpentier synthesized a group of compounds called penothiazines at the Specia Laboratories of Rhone-Poulinic near Paris. He was searching for a drug that would prevent surgical shock through its action as an antihistamine. In 1949, a French naval surgeon called Henri Laborit used one of these penothiazines for postoperative care of his patients and realized that it had unusual depressant effects on the central nervous system. Alerted to this observation in 1950, Simone Courvoisier of the Specia Laboratories screened the penothiazines for their depressant action and found that one of these, called promazine, was particularly effective. Charpentier went on to synthesize analogues of promazine and discovered a chlorinated derivative, which in the hands of Courvoisier was shown to have outstanding depressant actions on the central nervous system without any concomitant toxic effects. Laborit, now at the Val de Grace military hospital in Paris, used this to help patients in 1951, and realized that it had major effects in relaxing them before an operation. On 19 January 1952 he brought this observation to the notice of Joseph Hanson, director of neuropsychiatric services in the hospital. Hanson then successfully treated a manic patient who was soon calmed by the drug. They called the compound chlorpromazine and the Specia Laboratories marketed it as Largactil. Psychiatry was transformed overnight. Thousands were released from medical institutions. The physical restraints that had chained the mentally ill from time immemorial were removed from the inmates of the Paris asylums.

The next most significant discovery was the synthesis by Janssen of haloperidol. Janssen, working in his own company in Belgium in 1957, was attempting to improve on the known analgesics such as pethidine, and altered this into a butyrophenome compound. He noticed that besides its analgesic properties, the compound calmed excited mice in a way that was reminiscent of the actions of chlorpromazine. Janssen then set about synthesizing analogues of this compound so as to remove its analgesic properties but enhance its sedation or tranquillizing effects. In 1958, after preparing hundreds of derivatives, he came up with the most powerful tranquillizer ever discovered and named it haloperidol. This is about one hundred times more effective than chlorpromazine but has fewer side effects.

What are the side effects of these antipsychotic drugs, or neuroleptics as they are now called? The most obvious of them are movements characteristic of those with Parkinsonian disorders, such as tremor of the limbs. The suggestion that chlorpromazine and haloperidol are catecholamine receptor molecule blockers was a major breakthrough[10]. The idea was that the neuroleptics act at synapses by occupying sites, called receptors, on the membrane of neurons opposite nerve terminals which release substances like norepinephrine and dopamine; this action blocks the ability of these catecholamines to bind to the membrane receptors of the neurons during synaptic transmission, and so they are referred to as receptor molecule blockers. Anden and his colleagues finally showed in 1970 that the neuroleptics are dopamine receptor molecule blockers. This observation provided the key to the explanation of why the neuroleptics produce the Parkinsonian syndrome as a side effect to their antipsychotic action. The Parkinsonian condition arises as a consequence of the neuroleptics blocking the dopaminergic projection from the substantia nigra in the midbrain to the basal ganglia (Figure 5.6A and 5.6C), a condition that is known to give rise to the Parkinsonian syndrome. Most importantly, the various components of the schizophrenic syndrome enumerated above, such as hallucinations, avolitional changes and negative affect, that point to failure of neural networks in the cingulate cortex, frontal cortex as well as the amygdala and hippocampus, can all be considered to arise as a consequence of the excessive action of dopamine. This transmitter is released from the nerve terminals of the dopaminergic neurons in the ventral tegmentum that end in these parts of the cortex and hippocampus

(Figure 5.6A and 5.6B). The action of drugs, such as chlorpromazine and haloperidol, to block the action of dopamine released from the terminals of the ventral tegmentum then alleviates these symptoms of the schizophrenic syndrome. In addition, the ventral striatum of the basal ganglia projects to the cingulate cortex (Figure 5.6A). Excess dopaminergic transmission from the substantia nigra to the basal ganglia (Figure 5.6C) can lead to profound changes in the workings of the neural networks in the cingulate cortex, with all the implications for the pathological release of behavior associated with schizophrenia.

5.6 Dopamine receptor molecules and the alleviation of the distortions of consciousness

The dopamine hypothesis has centered attention on the different classes of dopamine receptor molecules that may be present in the brain and how these are distributed in the cortex. It is hoped that such an analysis will lead to the synthesis of drugs that act specifically on the particular class of dopamine receptor molecule that is mediating transmission at the site of the malfunctioning synapses. Determining the distribution of such receptor molecules will also allow identification of the neural networks in the cortex that are responsible for the schizophrenic condition.

Dopamine receptors are divided into two main classes: the D_1-like receptor class (includes D_{1a}, D_{1b} and D_5 receptors), which is defined as having a low affinity for the butyrophenomes such as haloperidol, and the D_2-like receptor class (includes D_{2a}, D_{2b}, D_3 and D_4 receptors), which has a high affinity for haloperidol. The D_1-like and D_2-like receptor classes are organized in a complementary fashion in some parts of the brain. For example, in the hippocampus the D_2-like receptors most densely populate the dentate gyrus and the adjacent subiculum whereas the D_1-like receptors are found in the entorhinal cortex and area CA1 of the hippocampus.

The diagrams in Figure 5.7 show identical sections through the regions of the hippocampus (indicated are the regions CA1, where the CA1 pyramidal neurons are found, and DG for the dentate gyrus) and the adjacent cortex (indicated are the regions SUB for subiculum, EC for entorhinal cortex, PR for perirhinal cortex, and the inferior temporal cortex). The top diagram in Figure 5.7A indicates the distribution of dopamine receptors (D1, D2), and the bottom diagram indicates the distribution of beta-adrenergic receptor molecules (β_1, β_2) in this region of the brain; Figure 5.7B shows the distribution of serotonin receptor molecules (5-HT_{1A}, 5-HT_2 in the top diagram) and the density of serotonin uptake sites located on serotonin nerve terminals (5-HT in the bottom diagram) in the inferior temporal cortex and the hippocampus.

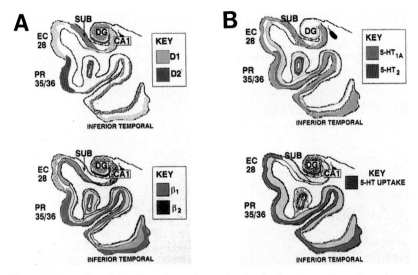

Figure 5.7   The distribution of receptor molecules for the transmitters dopamine, serotonin and norepinephrine, in the medial and inferior temporal lobe as well as the hippocampus.[11]

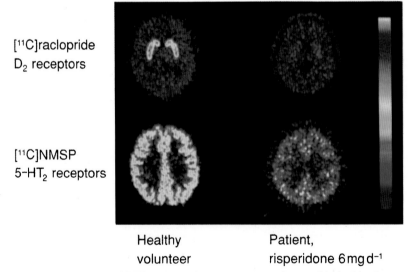

Figure 5.10   The use of PET to detect the receptor types that are blocked by the atypical neuropletic risperidone.[16]

There is great interest in the distinctions between the different classes of dopamine receptors, as in schizophrenics the D_2-like class is elevated but not the D_1-like class. Furthermore, it seems to be the D_4 receptors from the D_2-like receptor class that are elevated most dramatically, with the D_2 and D_3 receptors elevated by only 10% and the D_4 receptors by as much as 600%[12]. These increases occur throughout the temporal cortex, the associational cortex, and at the input to the hippocampus from the frontal cortex and the cingulate cortex, namely the dentate gyrus, as well as at the output of the hippocampus, namely the subiculum. Figure 5.8 shows the altered expression of receptor molecules for the transmitters dopamine, serotonin and norepinephrine in the temporal lobe and hippocampus of a patient with schizophrenia. The sites at which neuroleptics could act are indicated (by the asterisks) for the dopamine D_2 receptor molecule; the sites at which dopamine might interact with serotonin receptor molecules as well as with the class of β_2 norepinephrine receptor molecules are also indicated (by the boxes). The arrows show the direction of information flow through the synapses: the entire cortex has connections with the entorhinal cortex so that cortical information can flow from there to the dentate gyrus (DG) of the hippocampus via the perforant pathway. One of these inputs to the entorhinal cortex is from the temporal cortex, indicated, which shows an increase in the

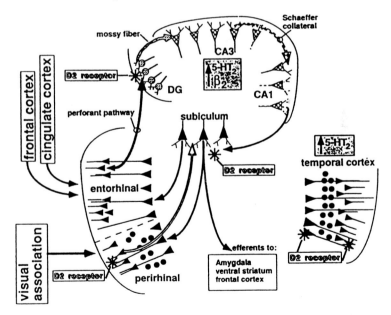

Figure 5.8   The altered expression of receptor molecules for the transmitters dopamine, serotonin and norepinephrine in the temporal lobe and hippocampus of a patient with schizophrenia.[11]

number of 5-HT_2 receptor molecules in schizophrenics (as indicated by the upward arrow in the box labelled 5-HT_2 in the diagram). In the hippocampus itself there is an increase in the expression of 5-HT_2 receptor molecules and of the β_2 receptor molecule type in schizophrenics (see the box labelled 5-HT_2 and β_2 with the upward arrow just above the subiculum) in the region where CA3 and CA1 pyramidal neurons are found. It is, therefore, in these regions of the cortex that the pathology of schizophrenia seems to be paramount[13] and, as shown in Figure 5.6B, it is just these regions that receive a dense dopaminergic projection from the ventral tegmentum.

The elevation of the D_4 receptors in the schizophrenic brain has provided a number of insights. Such receptors are most likely to be localized on nerve terminals themselves, rather than on the membrane of neurons on which nerve terminals abut. Figure 5.9 is a diagram of synaptic connections in the thalamus, one of the regions of the brain that consistently comes up in PET and fMRI studies as possessing abnormal synaptic function in the brains of schizophrenics. The spiny process of a dendrite (SP) protrudes from the main shaft of the dendrite (MS) of a neuron in the thalamus that mediates the visual input from the retina to the visual cortex (the cell body of this neuron is not shown at this level of magnification). Synaptic terminals (ST) from the retina form on excitatory synapses on this spiny process, and these terminals contain large numbers of round synaptic vesicles (sv) in which the synaptic transmitter glutamate is stored ready for release on arrival of a nerve impulse in the terminal; the dark bar at these terminals signifies the region at which this transmitter is released and acts. It will be observed that many of these synapses occur on other terminals, referred to here as inhibitory terminals (IT), with elliptical vesicles, which in turn synapse on the dendritic spine. These prevent excitation of the spine by the excitatory synapses.

As Figure 5.9 shows, there are nerve terminals that actually end on other nerve terminals, so that the release of transmitter from the latter terminals can be modulated by the release of transmitter from the former. This being the case, the release of dopamine from one terminal onto another terminal can alter the extent to which that terminal can effectively transmit information to the neuron on which it abuts. In this way dopamine may normally decrease the amount of transmitter released from a terminal on a neuron by acting on D_4 dopaminergic receptors on the terminal. If this terminal is normally inhibitory to the neuron (that is, it tends to stop the neuron firing impulses), then this is a form of disinhibition: we have the situation in which an inhibitory synapse is itself being inhibited, so allowing the neuron to be excited by other excitatory synaptic terminals more easily.

The D_4 dopamine receptor molecules bind the drug clozapine with high affinity. Whereas most neuroleptics act on the D_2 and D_3 receptors, clozapine acts

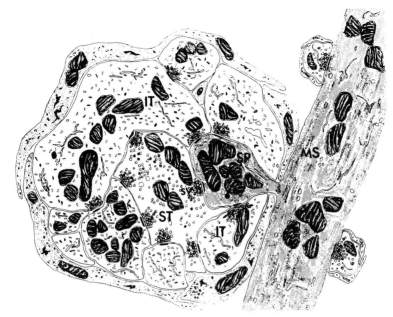

Figure 5.9   Diagram of synaptic connections in the thalamus.

almost exclusively on the D_4 receptors. This drug has been very successfully used to treat schizophrenic conditions that are opaque to the actions of the other neuroleptics, with about 30% of patients responding to the drug. Clozapine is referred to as an atypical antipsychotic because it does not elicit Parkinsonism. Many other neuroleptics produce the Parkinsonian syndrome as a consequence of their ability to block dopaminergic transmission from the substantia nigra to the caudate putamen of the basal ganglia (Figure 5.6C). This mimics the failure of dopaminergic transmission to the basal ganglia as occurs in Parkinsonism. The D_4 receptor is in relatively low amounts in the striatum of the basal ganglia compared with cortical regions and the hippocampus, as mentioned above. Clozapine therefore has relatively low binding in the basal ganglia, so that it does not produce the Parkinsonian syndrome.

Clozapine not only blocks D_4 dopaminergic receptors but also blocks receptors for the transmitter serotonin (5-HT), in particular the serotonergic receptor 5-HT_{2a}. The ability of clozapine to act as a powerful neuroleptic without the side effects of Parkinsonism may be due to its action on 5-HT_{2a} receptors as well as its high affinity for the D_4 receptors which are predominantly outside the striatum of the basal ganglia[14]. This combination of effects on both the dopamine D_4 receptors as well as the 5-HT_{2a} serotonin receptors has acted as a paradigm for the design of new atypical neuroleptics. Such a strategy, rather than using clozapine itself, has been necessitated by the discovery that clozapine produces bone marrow suppression (agranulocytosis), which necessitates continual haematological surveillance. The most successful of the new atypical neuroleptics is risperidone, which through an action on the D_2 and 5-HT$_2$ receptors is an improved antipsychotic. It is interesting in this regard that compounds that are specifically designed to block 5-HT$_2$ receptors, such as ritansevin, have no antipsychotic activity.

PET has helped establish the fact that clozapine and risperidone bind to both D_2-like and 5-HT$_2$ receptors, unlike many other neuroleptics[15]. The PET images in Figure 5.10 show the distribution of radioactivity in horizontal brain sections at the level of the caudate and the putamen (see Figure 5.6A). A control subject is shown in the two brain images on the left and a schizophrenic subject who had been treated with risperidone is shown in the two brain images on the right. The specific D_2 receptor blocker raclopride that had been made into a radioligand for the purposes of the imaging (therefore [^{11}C] raclopride) was used to determine the binding of risperidone to the D_2 receptors (upper row of images), and the specific 5-HT$_{2a}$ receptor blocker NMSP (made into the radioligand [^{11}C] NMSP) used to determine the binding of risperidone to the 5-HT$_{2a}$ receptors (lower row of images). It will be noted that the schizophrenic brain treated with risperidone does not show any binding of the specific D_2 and 5-HT$_{2a}$ blockers after the patient has taken risperidone, whereas the normal healthy brain does. This indicates that risperidone is binding to the D_2 and 5-HT$_{2a}$ receptors in the schizophrenic brain.

The atypical nature of the action of clozapine may therefore lie partly in its ability to block 5-HT$_2$ receptors (refer to Figure 5.7 for the distribution of 5-HT$_2$ receptors in the case of the temporal cortex and the hippocampus). These receptors appear to be located on dopaminergic neurons, so that they modulate the pattern of excitability of these nerve cells. The antipsychotic action of clozapine may then result from its action on the D_4 receptors belonging to the dopamine D_2-like class of receptor molecules (as well as in some cases on the serotonin 5-HT$_{2a}$ receptor molecules), and so modify the very synaptic pathways that are altered in the schizophrenic condition, as shown in Figure 5.8. The development of neuroleptics that will target both the D_4 dopamine receptor molecules and the 5-HT$_{2a}$ serotonin receptor molecules is therefore a main task of the pharmaceutical industry at this time. Indeed, as shown by Figure 5.11, an exponential increase has occurred in the extent of research on the most effective antipsychotic drug for schizophrenia, namely clozapine, and the most effective antidepression drug, fluoxetine (Prozac). Risperidone, another very effective antipsychotic drug for schizophrenics, has been synthesized only relativeïy recently but already shows a trend that indicates a future exponential growth in research on the action of this neuroleptic.

5.7 The site of the lesion in the schizophrenic brain and the distortion of consciousness

The discovery that the antipsychotic activity of the dopamine receptor blockers is in general proportional to their binding affinity to the dopamine D_2-like receptor molecules provided major support for the dopamine hypothesis. The more recent finding that the atypical neuroleptic clozapine (which has much greater antipsychotic activity than its binding to the D_2-like receptor molecules would lead one to expect) also acts as an antagonist to the serotonin 5-HT$_2$ receptor molecules has focussed attention on the distribution of the D_4 and 5-HT$_2$ receptor molecules in both the normal and schizophrenic brain. The elevation of the D_2-like and 5-HT$_2$ receptors in the temporal cortex and hippocampus, together with the elevation of the D_2-like receptors alone in the associational cortices, is consistent with PET studies pointing to these areas as possessing malfunctioning neural networks in the brains of schizophrenics. It is also consistent with a lesion in the dopaminergic projection from the ventral tegmentum to these areas as contributing in a major way to this psychotic condition.

As noted above, the thalamus shows up in imaging studies as abnormal in the brains of schizophrenics. This does not fit in neatly with the dopaminergic hypothesis, even when modified to include the serotoninergic receptors, as the thalamus does not receive a major dopaminergic projection from the midbrain. Furthermore, even though the D_4 dopaminergic receptor molecules are in relatively low numbers in the striatum of the basal ganglia, their numbers are elevated

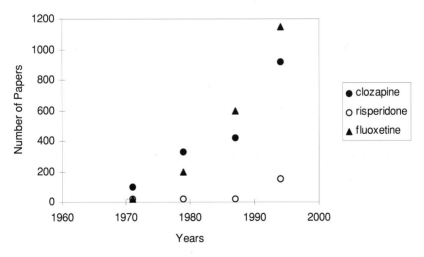

Figure 5.11   Changes in the rate of investigation of the major neuroleptic or antipsychotic drugs in the last three decades.[16]

two-fold in this region of the brain in schizophrenics[17]. The D_4 receptor is not synthesized in the striatum, but probably in the cell bodies of glutaminergic neurons from which the receptor is transported to terminals that release glutamate. The dopaminergic terminals from the substantia nigra may then release this transmitter onto the glutamate and D_4 receptor containing terminals, so inhibiting the release of glutamate from the terminals (Figure 5.9). An increase in the D_4 receptor in the striatum (comprised of the caudate nucleus, the putamen and the globus pallidus, and shown in Figure 5.6A) and the nucleus accumbens of the schizophrenic brain would then have the effect of inhibiting glutaminergic transmission in the striatum and nucleus accumbens, with the consequence that their connections with the hippocampus would malfunction. This sort of argument centers the schizophrenic lesion in the basal ganglion.

The modified dopamine hypothesis, which allows for the importance of serotonin in schizophrenia, has been applied to those areas of the brain that receive a rich dopaminergic projection from the ventral tegmentum. Whether this will provide an explanation at the molecular level for the distortions of consciousness that arise in the schizophrenic brain is not clear at this time. Indeed, it is possible that problems with transmission in the thalamus or even in the basal ganglia are the principal causes of this dreadful malady. One further problem with the dopamine hypothesis is that the antipsychotics take some weeks to take an effect. This implies that a simple binding of the drugs to dopamine or serotonin receptors, which should be completed in a few hours following administration, is not the determinant of the time course of alleviation of the schizophrenic condition after taking the drugs. The time scale is more like that to be expected for the growth of new nerve terminals or at least for the synthesis of a protein. Given the difficulty of the task, it is reassuring that the technical means to unravel this problem, such as imaging the brain and molecular biology, are now at hand.

5.8 What does schizophrenia reveal about the origins of consciousness?

Schizophrenia is in some cases a genetic disease, in which the near relatives of a schizophrenic have a tenfold greater likelihood of becoming schizophrenic themselves than if they are not so related. For this reason, there has been much effort lately to try and correlate the response of clozapine responding and nonresponding schizophrenic patients with the dopamine D_4 receptor allele frequency; that is, the frequency of variants in the D_4 genetic locus. This has not been successful because of the polymorphic forms of the D_4 receptor. Nevertheless, the effort to use molecular biological approaches to probe the origins of schizophrenia is a very important task. Schizophrenia is a development problem in which neural circuits, particularly those involving dopaminergic synapses, are not correctly formed. Indeed, the fact that the onset of the condition is most common in the late teenage years has been related to the final development of synaptic connections in the frontal lobes at this time. An attack of viral encephalitis can lead to disturbance in the development of these connections, independent of any inherited condition, and lead to schizophrenia.

Stress arising from social interaction, or very often the lack of such interaction, during the development of neural networks in the higher parts of the brain can tip an individual with a genetic propensity to schizophrenia into the condition. However, the presentation of the different levels of aoalysis of the brain given in Figure 5.1 seems to assume that the connections of the brain are given by a genetic blueprint, the product of which is then to be researched by the neuroscientist. This is unlikely to be true as part of the synaptic connectivity of the mature brain arises as a consequence of the interaction of one brain with another; that is, from social interaction. Figure 5.1 did not include this as a level of analysis, and must therefore be extended as shown in Figure 5.12 to include the interaction between nervous systems as another important area of study for those who wish to understand the development of the mature nervous system. Social interaction, especially during childhood and adolescence, has a very powerful influence on the development of synaptic connections within the higher neural networks such as those in the frontal cortex. There is an inextricable mix between the genetic and environmental factors that lead to the correct workings of the brain. An inappropriate balance between these can lead to mental illness, as in schizophrenia. Thus the analysis of the nervous system requires all of the levels given by the boxes in Figure 5.12, which might be regarded as the reductionist approach, as well as consideration of how interaction with other humans leads to the formation of the mature nervous system.

The consciousness of a paranoid schizophrenic may be haunted by the voices of threatening individuals, sometimes accompanied by horrendous images and emotions of persecution. Seeing the agony of such a schizophrenic gradually disappear after the administration of an antipsychotic evokes a profound gratitude towards researchers who are responsible for the neuroleptics. Indeed, to have a child in a near catatonic state for ten years or more with schizophrenia, and then to have within weeks the re-establishment of social intercourse and of an almost normal emotional life following the administration of clozapine, is a near miracle. The capacity of such drugs that interfere with dopaminergic, and in some cases serotonergic, synaptic transmission to remove debilitating hallucinations from consciousness highlights the fact that consciousness is a product of the working of neural networks, and helps identify which networks may be implicated. These include the frontal cortex, the cingulate cortex and the amygdala, areas which are so far removed from the sensory input to the brain that at this stage we know very little about their functioning. The fact that in this chapter an

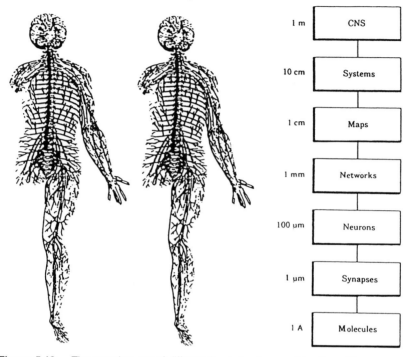

1 m	CNS
10 cm	Systems
1 cm	Maps
1 mm	Networks
100 um	Neurons
1 μm	Synapses
1 A	Molecules

Figure 5.12   The complete set of different levels for the consideration of human brain function.

identification has been made of some of the molecules involved in the malfunctioning of certain neural networks that give rise to hallucinations does not of course indicate that consciousness is the working of the networks. But it does offer persuasive argument that it will not be possible to progress far in the study of consciousness until an understanding is reached of how the distortions of consciousness in schizophrenia arise within neural networks through the molecular changes at synapses.

5.9 References

1.   Tootell, R.B.H., Reppas, J.B., Dale, A.M., Look, R.B. Sereno, M.I., Malach, R., Brady, R.J. and Rosen, B.R. (1995) Visual motion after-effect in human cortical area MT revealed by functional magnetic resonance imaging. *Nature*, **375**, 139–141.
2.   Miyashita, Y. (1995) How the brain creates imagery: projection to primary visual cortex. *Science*, **268**, 1719–1720.
3.   Roland, P.E. and Gulyas, B. (1994) Visual imagery and visual representation. *Trends in the Neurosciences*, **17**, 281–287.

4. Kosslyn, S.M. and Ochsner, K.M. (1994) In search of occipital activation during visual mental imagery. *Trends in the Neurosciences,* **17**, 290–292.
5. Kosslyn, S.M., Thompson, W.L, Kim, L.J. and Alpert, N.M. (1995) Topographical representation of mental images in primary visual cortex. *Nature,* **378**, 496–498.
6. Crick, F. and Koch, C. (1995) Are we aware of neural activity in primary visual cortex? *Nature,* **375**, 121–123.
7. Silbersweig, D.A., Stern, E., Frith, C., Cahill, C., Holmes, A., Grootoonk, S, Seaward, J., McKenna, P., Chua, S.E., Schnort, L., Jones T. and Frackowiak, R.S.J. (1995) A functional neuroanatomy of hallucinations in schizophrenia. *Nature,* **378**, 176–179.
8. Andreasen, N.C., Arndt, S., Swayze, V., O'Leary, D., Ehrhardt, J.C. and Yuh, W.T. (1994) Thalamic abnormalities in schizophrenia visualized through magnetic resonance image averaging. *Science,* **266**, 294–298.
9. Dolan, R.J., Fletcher, P., Frith, C.D., Friston, K.J., Frackowiak, R.S.J. and Grasby, P.M. (1995) Dopaminergic modulation of impaired cognitive activation in the anterior cingulate cortex in schizophrenia. *Nature,* **378**, 180–182.
10. Carlsson, A. and Lindqvist, M. (1963) Effect of chlorpromazine or haloperidol on formation of 3-methoxytyramine and normetanephrine in mouse brain. *Acta Pharmacologica et Toxicologica,* **20**, 140–144.
11. Joyce, J.N. (1993) The dopamine hypothesis of schizophrenia: limbic interactions with serotonin and norepinephrine. *Psychopharmacology,* **112**, S16–S34.
12. Seeman, P. and VanTol, H.H.M. (1994) Dopamine receptor pharmacology. *Trends in the Pharmacological Sciences,* **15**, 264–270.
13. Eastwood, S.L., McDonald, B., Burnet, P.W., Beckwith, J.P., Kerwin, R.W. and Harrison, G.P.J. (1995) Decreased expression of mRNAs encoding non-NMDA glutamate receptors GluR1 and GluR2 in medial temporal lobe neurones in schizophrenia. *Brain Research Molecular Biology,* **29**, 211–223.
14. Liebermann, J.A. (1993) Understanding the mechanism of action of atypical antipsychotic drugs. *British Journal of Psychiatry,* **163**, 7–18.
15. Strange, P.G. (1994) Dopamine D4 receptors: curiouser and curiouser. *Trends in the Pharmacological Sciences,* **15**, 317–319.
16. Farde, L. (1996) The advantage of using positron emission tomography in drug research. *Trends in the Neurosciences,* **19**, 211–214.
17. Murray, A.M., Hyde, T.M., Knobe, M.B., Herman, M.M., Bigelow, L.B., Carter, J.M., Weinberger, D.R. and Kleinmann, J.E. (1995) Distribution of putative D4 dopamine receptors in postmortem striatum from patients with schizophrenia. *Journal of Neuroscience,* **15**, 2186–2191.

CHAPTER 6   THE EVOLUTION OF CONSCIOUSNESS

6.1 Introduction

Is it possible to obtain information that indicates the stage in the evolution of the nervous system at which consciousness first appeared? Consciousness would seem to be a useful property to emerge from the workings of the nervous system if it confers an advantage in an evolutionary setting. Natural selection involves the interaction of a particular species with a particular environmental setting, as illustrated in Figure 6.1. In this case a particular phenotype (the largest shape in A) is favored for future reproduction (as shown in B, it matches the 'environment') with the subsequent loss of the remaining phenotypes (C). Consciousness may introduce a new element which permits a particular species to plan strategies which will allow favorable interactions in particular environments. The interaction is then no longer random and the species can enhance its chances of reproduction. It is not clear whether this capability of being able to plan strategies, which often involves the conferring on fellow members of the same species the property of a like consciousness, is unique to *Homo sapiens*. At what stage in the evolution of different species, shown in Figure 6.2, can it be said that consciousness is present? This problem was briefly considered in Chapter 4. The possibility of consciousness in the higher mammals (in particular macaque monkeys) is considered in this chapter, followed by the same consideration with respect to anuran amphibia (frogs). The techniques required to probe this question of consciousness in animals other than *Homo sapiens* are rapidly becoming available. This has occurred as a consequence of the development of PET and fMRI methods for pinpointing the areas of the brain that are active in any particular task that in *Homo sapiens* is known to involve consciousness (see Chapter 5). It has also been assisted by recent studies on the neural basis of attention and the solution of the binding problem (see Chapter 3). In addition, careful psychophysical studies of *Homo sapiens* with different lesions in the

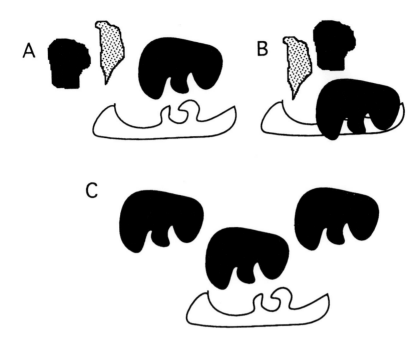

Figure 6.1   The usefulness of consciousness in an evolutionary setting.[1]

neocortex have recently been extended to other species. In this chapter humans will be referred to as *Homo sapiens* in order to attempt an objective evaluation of how the evolution of consciousness might be best investigated.

6.2 Evolution of the brain

Lower vertebrates possess brains that can be divided into three distinct regions that receive discrete sensory inputs (Figure 6.3). The first of these regions of grey matter (consisting of neuron cell bodies with their synaptic connections) is the forebrain, sometimes called the pallium or cortex, with the generic name telencephalon; this receives a direct olfactory input from the nose. Next is the midbrain, or mesencephalon, which receives visual input from the eyes. This is followed by the hindbrain which includes the cerebellum and has the generic name metencephalon; the hindbrain receives an input from the ears or the lateral line organs that can detect vibrations in the medium that the animal resides in. Note that the thalamus does not appear in Figure 6.3 as it is not a processing center for a particular sensory modality. Rather, the thalamus is a relay station for all sensory input to the major dorsal regions of grey matter with the exception of smell, which is dealt with by the olfactory bulb connected directly to the nose.

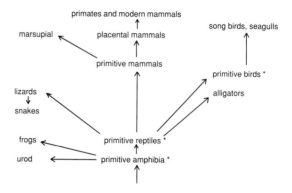

Figure 6.2   A simplified scheme of vertebrate evolution, from the primitive amphibia to the primates. The asterisks indicate extinct species.[2]

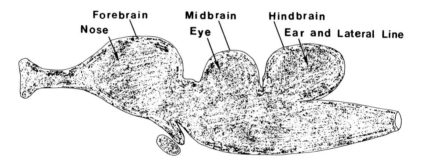

Figure 6.3   Relationship between the three major dorsal regions of grey matter in the brain of a lower vertebrate like a frog and the particular senses that they process.[2]

   The development of the three major dorsal and ventral regions of grey matter in the brain of higher vertebrates passes through a stage that can be related to the regions of grey matter found in the brains of adult lower vertebrates. At the developmental stage shown in Figure 6.4, the neural tube from which the grey matter develops is forming the ventricles, with the walls of the tube differentiating into grey matter that characterizes the collection of neurons and their synaptic connections in the principal divisions of the brain. These are the olfactory bulb, cortex or pallium (cloak), and basal ganglia of the telencephalon (TC); the thalamus and hypothalamus of the diencephalon (DC); the tectum of the mesencephalon (Mes C); and the cerebellum together with the medulla of the metencephalon

(Met C). The differentiation of the cortex or pallium of the telencephalon at successively higher levels of the evolutionary tree is shown in Figure 6.5 in which a lateral view of the cortex and the olfactory bulb is presented at different periods of evolution of a single hemisphere. At a primitive stage the entire hemisphere is given over to an olfactory bulb and a paleopallium that is concerned with smell (A). By the time of the amphibia the paleopallium differentiates on the dorsal side into the archipallium or hippocampus concerned with memory, and on the ventral side into the basal nuclei concerned, amongst other things, with motor activity (B). With the advent of the reptiles the basal nuclei have moved to the inner part of the hemisphere on the ventral aspect and so can no longer be seen in a lateral view (C). The neopallium or neocortex first appears in the advanced reptiles such as snakes, lizards and alligators (D). The archipallium or hippocampus is still evident in a lateral view of the brains of these species as is the paleopallium or smell part of the hemisphere. With the appearance of primitive mammals, the archipallium is transferred to the medial surface of the hemisphere and is therefore no longer observed in a lateral view; the neopallium has grown extensively in size although the paleopallium and olfactory bulb are still evident ventrally (E). The evolution of advanced mammals, including primates, involves a great increase in size of the neopallium with restriction of the olfactory bulb to an exclusively ventral part of the hemisphere (F). Thus the neopallium of different species of vertebrates varies over several orders of magnitude. As shown in Figure 6.6, the cortex is relatively smooth and tubular for lower vertebrates, such as frogs and snakes, with the extent of folding of the cortex into sulci and gyri increasing markedly after the evolution of the possum to reach its greatest extent in the primates.

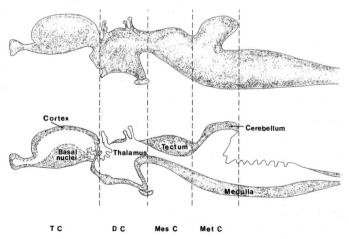

Figure 6.4   Development of the three major dorsal and ventral regions of grey matter in the brain of a higher vertebrate.[2]

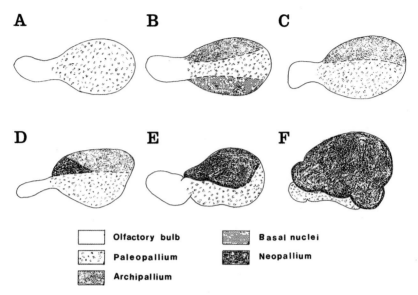

Olfactory bulb

Paleopallium

Archipallium

Basal nuclei

Neopallium

Figure 6.5   Differentiation of the cortex or pallium of the telencephalon at successively higher levels of the evolutionary tree.[2]

Since *Homo sapiens* are known to have consciousness, an inquiry into the evolution of consciousness is best begun by considering the primates closest to *Homo sapiens*. A number of species, each of different phylogenetic age, are available for study. Plesiadapis (upper left in Figure 6.7) was the common ancestor of the primates. It gave rise to the apes and to Ramapithecus in the Miocene epoch (12 million years ago). Ramapithecus gave rise to *Australopithecus robustus* and to *Homo habilis* in the Pliocene epoch (2 million years ago). *Homo habilis* then gave rise to *Homo erectus* in the Pleistocene epoch (1.6 million years ago) which in turn gave rise to *Homo sapiens* 400 thousand years ago. *Homo sapiens* has a brain weight of about 1350 grams but its ancestors, whilst possessing similar body weight, have a much smaller brain weight of about 590 grams. What evidence is there that primates other than *Homo sapiens* possess consciousness?

6.3 The site of visual consciousness in the primate brain: blindsight

There are different forms of 'blindness' following different lesions to the visual pathway from the retina to the cortex. Removal of the cortex gives rise to a state of cortical blindness in which there is an absence of what would normally be called conscious visual sensations. Qualia and their associated images are lost so that phenomenal vision is absent; that is, vision in which there is some kind of

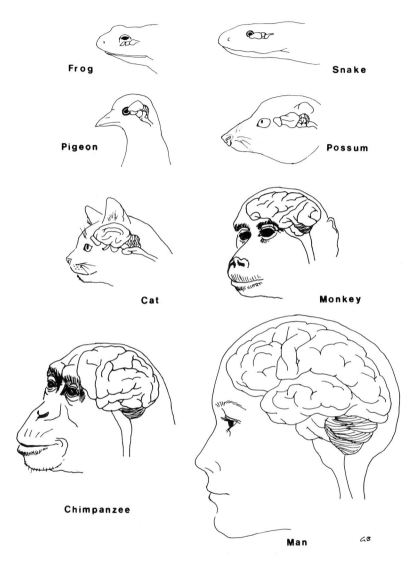

Figure 6.6   Comparative size of the pallium (cerebral cortex) of different species of vertebrates, drawn to scale.[3]

image made up of qualia consisting of regions of brightness and darkness and perhaps color. In the case of removal of just the striate cortex, that is V1, phenomenal vision is again absent. 'Blindsight' refers to the visual functions that remain after the removal of either the whole cortex or V1. Another form of 'blindness', called

apperceptive agnosia, which arises as a consequence of carbon dioxide poison-
ing that causes lesions in parts of the occipito-parietal-temporal cortex, is one in
which there is a failure to identify objects and shapes. In this case, an image is perceived so that phenomenal vision is present, but object vision is not. This results
in the failure to be able to draw, copy or match objects. A third form of 'blindness',
called associative agnosia, arises from lesions to the synaptic connections between the visual cortex outside the primary region (called the extrastriate visual
cortex) and the memory centers in the temporal lobe. Object vision is present, that
is the ability to bind qualia into objects and to see shapes, but there is a failure to
give these objects any meaning. There is then no appropriate use of objects and a
complete failure in the ability to be able to describe them. A final form of 'blindness'
that might be mentioned, called achromatopsia, involves lesions in the extrastriate
cortex which result in the loss of color so that the visual world appears in black and
white. The condition of blindsight is of particular interest in the context of seeking
the site, in the brain, of any form of visual consciousness.

The primary visual cortex has been discussed in Chapter 5 as being necessary for both visual perceptions and images in consciousness. Patients can still
make visual discriminations in the blind hemifield after lesions that remove V1 in

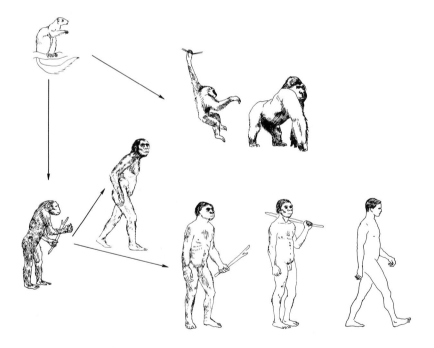

Figure 6.7   Evolutionary relationship (indicated by arrows) between *Homo
sapiens*, apes and monkeys.[3]

one hemisphere. In this case, visual discriminations can be made for objects that are observed by that part of the retina that projects to the topographically appropriate part of V1 which is lesioned. Such objects can be visually followed but not identified. This object detection is in general not accompanied by any awareness. However, if the object is moving, patients can sometimes report, in the absence of an experience of seeing an object, a contentless kind of awareness of something happening. Weiskrantz[4], who discovered blindsight, has studied patients who claim to sense 'a definite pinpoint of light' yet when questioned further comment that the experience does not 'actually look like a light but nothing at all'. Blindsight offers guidance as to which regions of the brain are involved in this kind of experience.

The kind of visual discriminations that patients can make in their blind hemifield after lesions to V1 in one hemisphere can be determined by experiments[5]. These test the extent to which the contrast of a stimulus can be discriminated for in a blind hemifield, with or without a residual form of awareness. Figure 6.8 depicts the results of such an experiment. Discrimination of whether an object was moving horizontally or vertically was made at different levels of contrast of the object. The luminance of the test stimulus was held constant, and background luminance in the blind hemifield was altered systematically, thus changing the contrast of the test stimulus. Subjects had to indicate (by guessing if necessary) whether the presented stimulus was moving horizontally or vertically by pressing one of two keys. In addition, subjects had to indicate whether they were aware or not aware of the stimulus by pressing one of two other keys. The line of filled circles indicates the percentage of trials in which the subject pressed the aware key. The line of open rectangles gives the percentage of trials when the subject pressed the unaware key and gave a correct response. The graph shows that there is a stable high level of performance independent of contrast if only those occasions in which the patient was unaware are counted. However, if only those occasions when the patient was aware are considered, then there is a steep decline with a decrease in contrast of the object as a consequence of an increase in the background luminance level. This experiment shows that detection of the stimulus is excellent in the absence of acknowledged awareness but deteriorates rapidly for acknowledged awareness. As the authors[4] of this research point out, it will be very interesting to examine brain activity by means of, say, PET scans, in the condition for which the patient reported awareness compared with the condition for which awareness was not reported. This may allow the areas of the brain that uniquely contribute to conscious perception to be identified. Results to this time suggest that this occurs in the prestriate cortex (V5), indicating that projections from the retina must reach this part of the visual cortex without going through the striate cortex (V1), a point further elaborated on below. The neural structures involved in consciousness might be identified

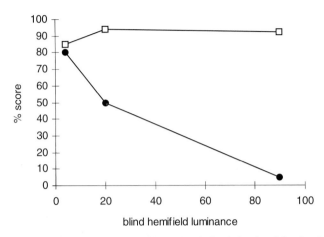

Figure 6.8   Conscious and unconscious visual discrimination following lesions to the primary visual cortex (V1).[5]

by combining a suitable experimental paradigm together with PET or fMRI on patients with blindsight.

Before attempting to see if it is likely that primates other than *Homo sapiens* experience blindsight (which will require primates without language to report claims of conscious experiences) it is instructive to consider the possibility of blindsight in infants. Infants prone to seizures may have their malformed hemispheres removed. This enables the investigation of possible blindsight in the absence of a hemisphere as well as in the absence of language with which to report the conscious concomitants of blindsight[6]. Magnetic resonance images are shown in Figure 6.9 of the brains of two infants before an operation in which each had a cerebral hemisphere, including both V1 and other parts of the visual cortex, removed. The operations were performed on the infants (who were prone to seizures) in their first year. The preoperative image of Figure 6.9a (made at the level of the thalamus and basal ganglia) shows a lack of both grey and white matter in the occipital region. The pre-operative image of the other infant (Figure 6.9b) is at the level of the basal ganglia and shows the abnormal left hemisphere; there is thickening of the cortex and an absence of gyri. Following such operations, single conspicuous targets in the half-field contralateral to the lesion could elicit fixations from the infants. These observations[6] point to the existence of blindsight in these infants, although whether there are any conscious elements associated with this is of course not known. One matter that does seem likely, however, is that with no hemisphere to mediate their visual responses, the infants' detection and orientation mechanisms must be mediated by subcortical systems.

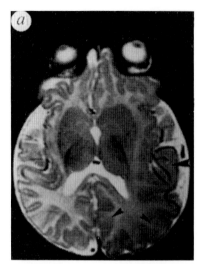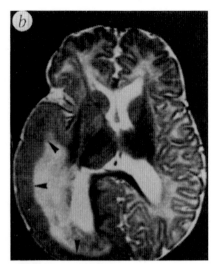

Figure 6.9 Blindsight in infants lacking one cerebral cortex.[6]

Monkeys that have lesions in their primary visual cortex can learn to detect, localize and distinguish between visual stimuli presented within their visual field defects. Patients with blindsight deny seeing a stimulus presented to them in their blind hemifield which they may nevertheless respond to. The question arises as to whether it is possible to design experiments that indicate if monkeys also have blindsight, even though they cannot verbalize as to whether they have consciousness of anything, and therefore what the contents of such a consciousness might be.

It is possible to determine if monkeys are also likely to have blindsight by testing if they see objects in their visual field defect. To this end, four monkeys were tested: one monkey was normal (Rosie) whilst the others (Dracula, Lennox and Wrinkle) were phenomenally blind in the right visual hemifield (due to the removal of V1 in the left cerebral hemisphere and the cutting of the corpus callosum)[7]. These monkeys showed very good detection in tasks for which a visual stimulus was presented on every trial. They were then tested in a series of trials in which half of them were blank; that is, no visual stimulus was presented. A diagram is presented in Figure 6.10A of the visual display used in the first part of the experiment. The fixation point (which the monkey must touch to begin the trial) is indicated by F; the other four rectangles indicate the positions of the visual square stimulus for testing the normal left visual hemifield (left two squares) and the right visual hemifield affected by the lesioned left cerebral hemisphere (right two squares). The experimental results[7] are given in histogram form. The first set show the correct responses of touching the activated visual square (and thereby receiving a reward). For each monkey, the first column is the

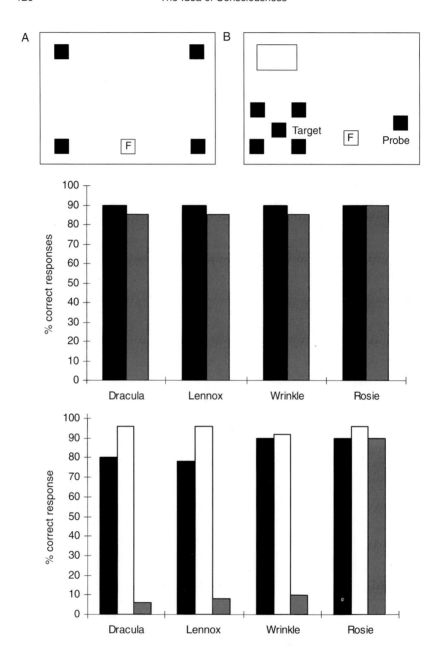

Figure 6.10   Tests for blindsight in monkeys.[7]

percentage of correct responses to the left stimulus and so tests the normal left hemifield. The other column is for the right stimulus. It will be noted that all the monkeys have a better than 90% response for both left and right visual fields, independent of whether they have a lesion or not. Thus if the monkeys devoid of a left primary visual cortex are 'forced' to produce a motor response (as a consequence of a food reward) for a visual clue in their phenomenally blind right hemifield, they will do so with a high degree of success. Figure 6.10B gives a diagram of the visual display when using blank trials and probes. As before, the monkey starts the trial by pressing F. On light trials the visual stimulus appeared transiently at one of the five positions shown in the bottom left quadrant and the monkey touched the position of the stimulus to receive a reward; since this occurred in the left half field, the monkeys had phenomenal knowledge of the image; that is, they saw it. On blank trials no visual stimulus was presented and the monkey had to touch the outlined rectangle that was permanently present in the upper left quadrant, as shown, for which they received a reward. On probe trials the visual stimulus appeared on the right at the position marked 'Probe', and so appeared in the phenomenally blind field. Since the monkeys received a reward if they correctly recorded a phenomenally observed blank falling in their left visual field, they were aware that food would also be provided if they correctly recorded a blank in their phenomenally blind right visual field. The second set of histograms gives the percentage correct responses for the four monkeys. For each monkey the first bar indicates the percentage that were correct when a visual stimulus appeared in the left hemifield; the second bar indicates the percentage correct on blank trials, when no stimulus was presented; the third bar shows the percentage correct for one hundred probe trials in which a visual stimulus, known from the initial experiments to be well above detection threshold, appeared in the right hemifield. All monkeys made some errors, even in the normal visual field. The results show that monkeys that were lesioned recorded no phenomenal responses in their right visual hemifield, indicating that they had blindsight[4]. That is, they were able to use this hemifield for motor acts that received rewards but it was without phenomenal content.

Can the area of the brain that is responsible for blindsight be identified? In *Homo sapiens*, as already mentioned, blindsight may be mediated by pathways to the neocortex from the retina that do not go through the geniculate in the thalamus and therefore do not go through V1. Alternatively, it may be mediated by pathways that involve the tectum or superior colliculus of the mesencephalon; that is, through subcortical systems. It seems likely that the infants who possess only one hemisphere do possess blindsight, so that their detection and orienting abilities must be subserved by subcortical systems such as the superior colliculus, which is called the optic tectum in some vertebrates, although other subcortical systems might also be involved[8]. In the case of adult patients with damage to V1 in one hemisphere, both subcortical neural systems as well as cortical systems that receive inputs from other parts of the thalamus that do not use the geniculate projection to V1 are likely to be involved. PET studies have been used to determine if signals from the retina reach V5 in the neocortex, responsible for detecting visual motion, in a patient with lesions to V1. The activity in V5 (that is, MT) associated with the waterfall effect has already been described in Chapter 5. As mentioned earlier in this section, Weiskrantz[4] has studied patients who claim to sense 'a definite pinpoint of light' which 'looks like nothing at all'; that is, they have a contentless form of awareness. These patients were consciously aware of a moving target as well as the direction of movement of the target without being able to identify the target, according to their verbal reports. Their PET scans showed increased activity in V5 during the task, indicating that this is the area responsible for both the detection of the moving object and the conscious awareness of the movement. They were there-fore aware of the moving target in their blindfield and able to report verbally its direction of motion. Thus there must be a subcortical input to V5 with informa-tion from the retina that allows V5 to compute the direction of motion of the object without having any input from V1. Most importantly, from the point of view of the present considerations, V5 can contribute directly to the conscious experience of visual motion[10].

There are other examples of patients with lesions to the neocortex who suffer from a form of apperceptive agnosia as they are able to manipulate objects without having any visual consciousness of at least some of the objects. These amount to rather complex cases of a kind of blindsight. The interesting case of a brain-damaged patient who could see only one object at a time, and whose ability to complete actions involving two objects in central vision depended on the semantic or functional relationship between the objects, has recently been described[9]. This patient suffered a left-hemisphere stroke which eventually gave rise to bilateral lesions that did not involve visual defects. Thus the patient could not see in consciousness two objects at the same time, only one, but could nevertheless manipulate two if they were of like semantic content. There must have been some solution to the binding problem in order for the patient to be able to manipulate both objects normally, yet only one of the objects was in visual consciousness as the patient did not report sensations associated with awareness of the second object. The patient reported that[9] 'As my pen ap-proaches my grandson's scribbles, the tip disappears and I cannot draw. However, I can write my name'. 'I see only the bottle or its lid but can screw on the lid; only the toothbrush or the tube but can squeeze out the toothpaste.' When presented with two pictures of objects the patient immediately reported seeing only one. Figure 6.11 illustrates an experiment that highlights this schism between the ability to manipulate objects of different semantic content and their presence in visual consciousness[9]. This is a pointing task in which the patient was asked how many pictures of objects she could see, and upon replying was asked to point to them. Two pictures pasted on cards were presented side by side in central view. The semantic relationship between the cards was manipulated. Shown in the upper part of the figure, the patient points to both (apple and orange) yet reports seeing only one. This always occurred for pictures and objects from this category (she often asked how she could point to two when she saw only one). In the lower part of the figure there is no semantic relationship. In this case she points to and reports seeing only one card (monkey or apple). The results of this study suggest that the motor system receives instructions from the visual system that do not pass through consciousness and are akin to blindsight.

Blindsight has many aspects to it that need clarification. One of these concerns whether there is some aspect of conscious visual experience associated with the workings of subcortical processes[6, 11] like those that involve the superior colliculus of the mesencephalon. If there is such experience, then species with homologues of these subcortical areas that play a major role in their visually dependent behavior may also have consciousness associated with the workings of these parts of the brain. Of course this might not be the case, but it would be unwise to assume that such neural areas were without visual consciousness in these other species, as is argued more fully below. On the other hand, given the importance of the superior colliculus and optic tectum in the control of muscles such as the extraocular muscles involved in responses to visual stimuli, it may be that these neural structures give rise to a consciousness of movement of the kind mentioned in Chapter 4.

6.4 At what point during evolution of the brain do synchronized 40 Hz oscillations first appear in the telencephalon?

It was argued in Chapter 3 that the appearance of coherent 40 Hz oscillations in different parts of the neocortex may provide conjunctive properties for solving the binding problem in which different features of an object that are processed in different parts of the neocortex are bound together to give a unified object that may be perceived; that is, may arise in consciousness. However, attentional mechanisms must also operate to sculpt out of the large number of objects for which the binding problem is solved at any given time just those that will enter consciousness. According to this scenario, it is necessary for synchronous 40 Hz oscillations to be generated in order for the binding problem to be solved so that consciousness can have coherent visual content. What species possess such 40 Hz oscillations? The answer to this question will not tell us which species possess visual consciousness for it has already been argued that the

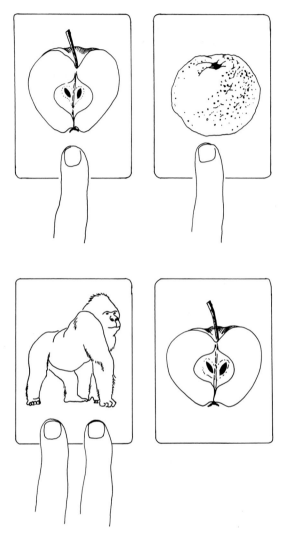

Figure 6.11   The ability to complete actions involwing two objects in central vision is influenced by the semantic or functional relationship between the objects in a brain-damaged patient who can see only one object at a time.[9]

oscillations are merely a requirement for a coherent visual perception, but are not necessary for the perception itself. It seems likely that the telencephalon of most species will have to possess such a mechanism for solving the binding problem so that the animals can navigate in the world. Coherent 40 Hz oscillations can be detected in the magneto-encephalogram in awake humans and during rapid eye movement sleep associated with dreaming. Spectral analysis of the magneto-encephalogram of awake and unrestrained rats shows activity is higher in about the 40 Hz range than it is in rats that are in non-rapid eye movement sleep[12]. This suggests that the rat cortex in the awake state is solving binding problems. Such 40 Hz oscillations also occur in rat thalamic reticular neurons[13] and even in slices of rat neocortex[8]. It has already been shown in Chapter 3 that these oscillations occur in the neocortex of cats. However, there has been no systematic study of the species dependence of these oscillations.

Although the emphasis has been placed on the role of 40 Hz coherent oscillations in solving the binding problem both here and in Chapter 3, this might not be the only mechanism by which this occurs. The binding problem may also be solved by using the time of arrival of impulses at a particular recognition neuron in the neocortex. Hopfield[14] has suggested that the relative timing of impulses between cortical neurons may give information about different components that have to be brought together in order to solve the binding problem and so have an holistic experience of, say, an object. In this case, the binding problem is solved when the impulses arrive simultaneously at a 'recognition neuron'. Evidence for this idea has recently been provided[15] by showing that the relative impulse timing of neurons in the neocortex, rather than their rate of firing or their characteristic of being phase locked to a stimulus, is probably a process for signalling object features. In this case, if a number of temporally coordinated impulses in a particular part of the neocortex converge on a particular neuron, then the firing of this neuron may contribute a solution to the binding problem. It is then necessary to measure impulse firing in individual neurons of a population to ascertain their relative impulse firing in order to see if the binding problem is being solved. This may be a formidable task if it is to be carried out routinely.

6.5 At what point during evolution of the brain do PET and fMRI studies indicate psychophysical parallels with *Homo sapiens*?

The fact that either PET or fMRI can detect the regions of the brain that are active during perception, as described in Chapter 5, does not mean that this activity gives rise to consciousness. It does indicate that these neurons are carrying out computations that may be associated with a particular act of perception. The cases of blindsight show how objects can be manipulated by a patient who has no, or very little, phenomenal perception of the object in consciousness. However, the use of PET and fMRI techniques does identify the active regions of the brain at any one time, some of which give rise to consciousness if it should occur during any particular act of perception. This was illustrated with regard to the waterfall illusion by showing it is located in V5 of the neocortex (see Chapter 5). There certainly may be a multiplicity of sites that are found to be active using PET and fMRI techniques, with only one or even none of these giving rise to the content of consciousness. Such imaging of the brains of species other than *Homo sapiens* is then a first requirement for identifying an area of the brain that might give rise to the content of consciousness in a particular act of perception by that species. Furthermore, it is possible to set up experimental paradigms, like that for testing the neorcortical site of origin of the waterfall illusion in humans, and then applying the same paradigm to other primates, such as macaques. In this way it might be possible to infer whether V5 in the monkey is also responsible for a waterfall illusion in the consciousness of that species. Of course, the waterfall illusion experiment required that the human subject report on the rate of disappearance of the illusion. Monkeys can also be trained do this through appropriate reward strategies based on the assumption that they are also experiencing a waterfall effect illusion under the appropriate experimental conditions. Such behavioral training will be difficult for lower mammals, but it would at that stage be very interesting to see if V5 was also uniquely active in their neocortex during exposure to the same experimental conditions that give rise to the waterfall effect in *Homo sapiens*.

This use of imaging techniques on different species is now becoming available. A system for carrying out PET studies on conscious monkeys has been devised[11] and used to show that there are topographically organized maps of the monkey's body surface in the somatosensory cortex of these species. High resolution PET is now being used on cats to determine local blood flow and therefore neuronal activity in different parts of the neocortex under different experimental conditions[16]. Furthermore, high resolution PET can now be used on small animals such as rats[17], enabling the binding of radioactive compounds in different parts of the rat brain to be monitored[18] at a level of resolution that is less than that of the size of the somatosensory cortex[19]. These technical developments mean that the comparative study of the areas of the brain involved in perception and other activities is now ripe for investigation. This can be carried out within the context of the appropriate design of experiments to elucidate which, if any, of these areas that are active contribute to consciousness.

6.6 Is there visual consciousness in amphibia?

The philosopher Dennett has commented that[1]:

> In this case, the toad's status falls to 'mere automaton', something that we may interfere with in any imaginable way with no moral compunction whatever. Given what we *already* know about toads, does it seem plausible that there could be some *heretofore unimagined* feature the discovery of which could justify this enormous difference in our attitude? Of course, if we discovered that toads were really tiny human beings

trapped in toad bodies, like the prince in the fairy tale, we would immediately have grounds for the utmost solicitude, for we would know that
in spite of all behavioral appearances, toads *were* capable of enduring
all the tortures and anxieties we consider so important in our own cases.
But we already know that a toad is no such thing. We are being asked to
imagine that there is some x that is nothing at all like being a human
prince trapped in a toad skin, but is nevertheless morally compelling.

Dennett is assuming here that toads do not have consciousness but are[1] 'rather an
exquisitely complex living thing capable of a staggering variety of self-protective
activities in the furtherance of its preordained task of making more generations of
toads'. On the other hand the philosopher Nagel in a discussion of *What is it like
to be a bat?* comments that[20] 'Conscious experience is a widespread phenomenon.
It occurs at many levels of animal life, though we cannot be sure of its presence in
the simpler organisms, and it is very difficult to say in general what provides evidence of it. (Some extremists have been prepared to deny it even of mammals other
than man.) No doubt it occurs in countless forms unimaginable to us ...' It is not
clear that either of these extreme views is correct. Nagel goes on to say that[20] 'Bats,
although more closely related to us than those other species, nevertheless present
a range of activity and a sensory apparatus so different from ours that the problem
I want to pose is exceptionally vivid (though it certainly could be raised with other
species) ... We must consider whether any method will permit us to extrapolate to
the inner life of the bat from our own case, and if not, what alternative methods
there may be for understanding the notion ... I want to know what it is like for a *bat*
to be a bat'. Nagel has proposed a project here that is so daunting in complexity as
to deter any approach to the problem of animal consciousness. However, in the
present section it is argued that it might be possible to find a convincing way to
show experimentally that even the humble amphibia, assumed by Dennett to be
automatons, do or do not have some form of consciousness. It is necessary to
break the problem down into an inquiry of how the best analyzed forms of consciousness in *Homo sapiens*, such as those involved in perception, may or may
not occur in frogs.

A comparison of the brain of frogs with that of other species shows the
relatively small size of the cerebral cortex of the telencephalon, the optic tectum
of the mesencephalon and the cerebellum of the metencephalon (Figure 6.12).
The optic tectum of the frog is of special interest in any studies concerning the
possibility of consciousness in visual perception as this is the dominant visual
processing part of the frog's brain, which is comparatively large when compared
with the rest of its brain. The optic tectum is evident for all species considered
except that of *Homo sapiens*, in which case the mesencephalon is overlaid by
the cerebral cortex and so the tectum is not present in a lateral view. The principal

nerve pathways mediating vision in the brain of these different species, including the frog, are illustrated in Figure 6.13. Shown is the projection from the retina to the geniculate (G) and pulvinar (R) of the thalamus in the diencephalon and to the tectum in the mesencephalon. Projections may then occur from these to the telencephalon, the laminated general cortex (Gc), the dorsal ventricular ridge (DVR), the hyperstriatum (Hyp), the ectostriatum (Ect) and to areas 17, 18 and 19 of the cortex depending on the species. In cats, the retina projects to the geniculate as well as the tectum with the latter connecting to the pulvinar; both the geniculate

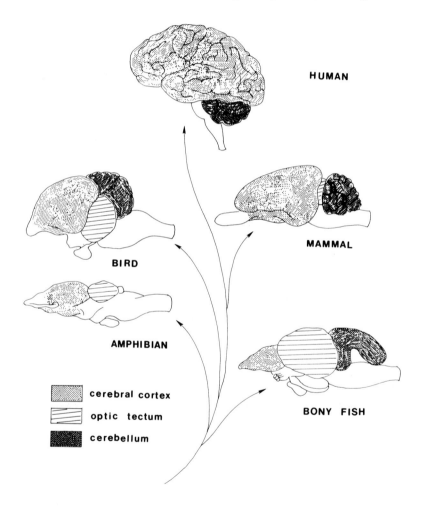

Figure 6.12　Lateral view of a single hemisphere of the brains of a number of species. These have not been drawn to scale.[21]

and the pulvinar then connect to the visual cortex of the telencephalon, with the geniculate connecting to V1. In birds, the retina again projects to the geniculate and the tectum, with the latter connecting to the pulvinar; the geniculate then projects to the hyperstriatum and the pulvinar to the ectostriatum, both of these structures of telencephalic origin. In frogs, the retina projects to the geniculate body and to the tectum which in turn projects to the geniculate; this geniculate body connects to a non-laminated telencephalon. The details of this pathway from the retina to the telencephalon in the frog are shown in Figure 6.14. One retinal projection is to the optic tectum (Ot) and from there to the posterior lateral nucleus of the thalamus (pl) which in turn projects to the lateral forebrain bundle of the telencephalon (Lfb). The other retinal projection is to the anterior posterocentral nucleus of the thalamus (Pca) and from there to the medial forebrain bundle of the telencephalon (Mfb) and the medial pallium of the telencephalon (mp). These comparisons show that the frog's telencephalon and optic tectum do receive inputs from the retina, as is the case for the higher

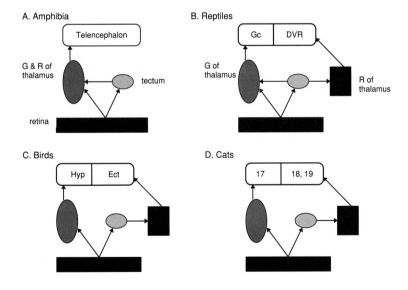

Figure 6.13   Highly simplified diagrams of the principal nerve pathways mediating vision in the brain of a number of different species.[3]

mammals including *Homo sapiens*. It is, therefore, not possible to justify the claim that visual consciousness cannot be present in frogs, as it is likely to be the function of the telencephalon. Of course the telencephalon of the frog is very different to that of mammals, so there is no suggestion that a telencephalon per se is capable of generating consciousness.

Figure 6.15 shows a frog's brain in dorsal (C) and lateral (B) view, together with a diagram (A) of the projection of the optic tract to the principal areas of the brain subserving vision. The large size of the optic tectum of the frog, compared with that of the cerebral hemispheres (Figures 6.15A, B and C), reflects the fact that it receives the entire retinal input to the brain, whereas in mammals the optic tectum, or superior colliculus as it is called for these species, receives only part of the retinal projection. It is therefore appropriate to inquire as to whether frogs are conscious of a visual perception in the tectum rather than the telencephalon. This optic tectum has been a source of modelling of neural networks for the reflex control of eye position in the frog. From the tectum, projections occur to motor neurons that control the left medial and right lateral rectus muscles of the

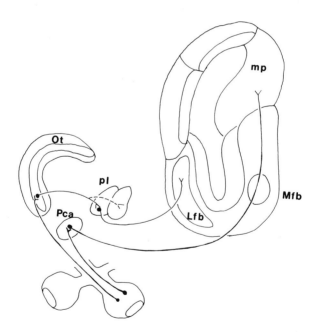

Figure 6.14   Details of the retinal projection to the optic tectum and the telencephalon of the frog.[22]

eyes as well as to those controlling vertical movement. This tectum has also been subjected to a control system analysis of the process of eye movement involved in saccades (Figure 6.16). According to this theory, a retinotopic mapper occurs in the tectum which projects to a saccade burst generator in the brain stem. This then receives a code for target position as determined by the position of a peak of activity in a neural map. However, all of these considerations of the functioning of the tectum emphasize Dennett's claim[1] that 'the toad's status falls to "mere automaton" '. The question being asked is: does the tectum give rise to consciousness?

One clue on how to approach this problem is offered by Nagel in his comment that[20] 'It may be easier than I suppose to transcend interspecies barriers with the aid of the imagination. For example, blind people are able to detect objects near them by a form of sonar, using vocal clicks or taps of a cane. Perhaps if one knew what that was like, one could by extension imagine roughly what it was like to possess the much more refined sonar of a bat'. Following this idea, is it possible to determine if there is consciousness of visual perception in the superior colliculus of human patients who have only one hemisphere? Even given that the superior colliculus of *Homo sapiens* and the optic tectum of frogs are not identical in terms of their neuronal connectivity, they do subserve many of

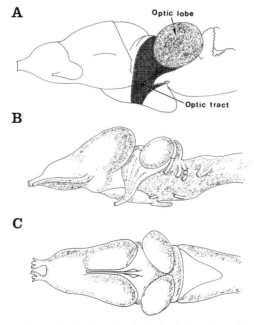

Figure 6.15   Lateral and dorsal view of a frog's brain together with a diagram of the projection of the optic tract to the principal areas of the brain subserving vision.[13,23]

the same functions. If it should be proved that there is no visual or motor consciousness in the superior colliculus of *Homo sapiens*, then this would make it unlikely that there is such consciousness in the optic tectum of frogs. It is necessary then to first ascertain whether blindsight in patients without one hemisphere does give rise to a contentless kind of awareness in consciousness, using suitable psychophysical experimentation of the kind already described in this chapter. Imaging techniques are then required to determine the site or sites in the brain that are active during this experience, which for the sake of argument we will claim is in the superior colliculus of the mesencephalon. A check is next made with magneto-encephalographic studies that 40 Hz oscillations are associated with this experience arising from the colliculus. If the patient is conscious of the object, then the binding problem must have been solved. Therefore, there must exist 40 Hz oscillations arising from the superior colliculus and PET will show high levels of activity in this region.

This experimental approach now requires us to go to another primate species, such as the macaque monkey studied in the experiments illustrated in Figure 6.10 that had lesions to its neocortex, and carry out all of the procedures enumerated above for the human patient. If the experimental results of these investigations are the same as that for the human patient, then it is likely that there is some visual consciousness in the superior colliculus of the macaque. It has already been shown that in the research on the macaque illustrated in Figure 6.10 it is possible to design experimental procedures that give information about the probable contents of consciousness of the monkey by analogy with the human situation. If all the experimentally measured concomitants of a consciousness in part of the brain of *Homo sapiens* are also present in the same part of the brain of a macaque, then it is reasonable to infer that consciousness is also present in that brain region. If the results of the experiments on primates show that it is likely that consciousness does exist in the superior colliculus, then it is worthwhile extending this process to other mammals, and if that should also prove

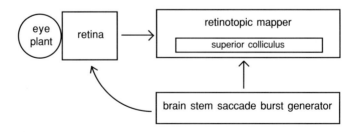

Figure 6.16   Control system analysis of the process of eye movement involved in saccades.[24]

positive with respect to these various tests for consciousness, then we may proceed to the lower vertebrates and to frogs. It is not argued here that birds will fail the test for consciousness in their optic tectum if certain mammals, such as the possum, fail the tests. But a gradualist's approach through different species down the evolutionary tree is required. A key point is that if *Homo sapiens* fail to show such consciousness in the superior colliculus, then the link to the possibility of consciousness in this structure in other species is lost, although it would still certainly be interesting to carry out the array of experiments suggested above on other species. Simply testing out the possibility that the superior colliculus possesses consciousness is a large research program. It is a research program that can be extended to consider which other parts of the brain possess a consciousness in both *Homo sapiens* and other species.

Arguing from experimentation that consciousness of a particular content occurs in a specific part of the brain of another species does not seem to be too far fetched when talking about primates other than ourselves, or indeed perhaps about other mammals. But what of frogs? The case for frog consciousness was taken up in this chapter in order to be deliberately provocative. I hope that the reader has been stirred to the realization that there is no evidence which disproves that frogs have a level of consciousness in their comparatively large optic tectums of the kind that may exist in the superior colliculus of a blindsighted patient with no hemisphere. Here it is suggested that it is now open to experiment as to whether frogs are automotons or whether some parts of their brains give rise to consciousness. This may hardly compare with the richness of our own consciousness, more like that of blindsighted humans in the case of visual perception. Nevertheless, it may be present in a form that we can relate to our own consciousness in a way that Nagel originally suggested.

6.7 Evolution of the emotional centers of the brain

In this chapter the possibility of consciousness in the brains of species other than *Homo sapiens* has been restricted to that of visual perception. Indeed, visual perception has dominated this book in the search for the biological origins of consciousness. This is, of course, necessitated by the fact that it is the visual system that is most readily available for experimentation in relating the contents of consciousness to the neuroscience of the brain. But how do all the other attributes arise that contribute to the contents of consciousness in a normal brain? Perhaps at the opposite extreme to the apparently tractable problem of how visual perception arises in consciousness is the question of how emotional states in consciousness occur. Indeed, is the capacity to feel joy and horror restricted to our own species or are there some versions of these emotions in the consciousness of other species? Do monkeys or possums or frogs have some version of feeling in some kind of a consciousness?

Damasio[25] has argued that the origin of emotions and emotional memories resides in the amygdala, in close relation to the declarative or explicit memory systems in the hippocampus. The amygdala is then responsible for emotional memory; that is, for the feelings of sadness, horror, joy, et cetera, that often accompany declarative memories that are processed by the hippocampus. The schematics in Figure 6.17 illustrate this point of view. The black perimeter in each represents the brain and the brainstem. A indicates the amygdala, H the

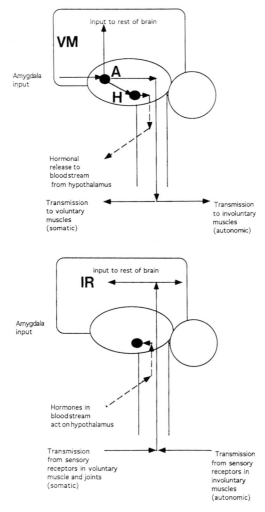

Figure 6.17   The origins of emotion reside in the amygdala.[25]

hypothalamus, IR the internal responses and VM the frontal cortices. In Damasio's scheme a particular stimulus, such as a specific smell or sound, activates the amygdala, giving rise to a number of different responses (upper schematic). These mostly involve changes to the internal organs such as the viscera and the vasculature, parts of the autonomic nervous system concerned with maintaining the homeostasis of the body, as well the release of chemical substances into the bloodstream from the hypothalamus. These substances from the hypothalamus in turn act on the internal organs. In some instances there are also changes to the muscles of the face and limbs. The sensory receptors in the internal organs of the viscera and vasculature, as well as in muscles of the limbs and face, then deliver information to the neocortex as to the contractile state of these organs (lower schematic). This is then read as a particular emotion, such as joy or horror, by the cortex and the emotion may then enter consciousness.

In order to make this process of emotional memory clearer, consider the following example. At a young age you are involved in a serious accident, in which cars are smashed and the scene is permeated with the sound of their jammed horns. Your parents are severely injured but you are unharmed. The scene of the accident is horrific, amounting to a traumatic experience for you that is accompanied by changes in blood pressure due to an accelerated heart beat and constriction of blood vessels, as well as in marked changes in the activity of the gastrointestinal tract and the urinary bladder. The sensory receptors in these organs pass information concerning the contractile state of the organs to the amygdala, and it places the temporal relationship between the sound of the horns and the condition of the viscera and vasculature into memory. Many years later, when you are about to walk across an intersection, a car blasts its horn which happens to be at the same pitch as that which was heard at the time of the accident. This stimulates the emotional memory in the amygdala, which then activates the viscera and vasculature in the same pattern of contraction or relaxation as that which occurred at the time of the accident. Sensory receptors in these organs send information concerning their changed condition to the neocortex, where they give rise to the sensation of horror in consciousness. Whether this emotional condition is accompanied by explicit memories of the accident is dependent on parts of the brain other than the amygdala, so that it is possible for the emotional memory to be triggered without a concomitant explicit memory. In this case, a traumatic emotional experience may occur to you without there being any apparent reason for it to occur. Clearly the problem of how emotions arise in consciousness is one that is far more difficult than that of how visual perception arises. There is much research to be done here in order to bring emotional consciousness within experimental analysis.

6.8 References

1.   Dennett, D.C. (1996) *Kinds of Minds*. New York: Basic Books. Harper Collins.
2.   Romer, A.S. (1960) *The Vertebrate Body*. London: W.B. Saunders and Co.
3.   Sarnat, H.B. and Netsky, M.G. (1981) *Evolution of the Nervous System*. New York: Oxford University Press.
4.   Weiskrantz, L. (1986) *Blindsight: A Case Study and Implications*. Oxford: Oxford University Press.
5.   Weiskrantz, L., Barbur, J.L. and Sahraie, A. (1995) Parameters affecting conscious versus unconscious visual discrimination with damage to the visual cortex (V1). *Proceedings of the National Academy of Science of USA*, **92**, 6122–6126.
6.   Braddick, O., Atkinson, J., Hood, B., Harkness, W., Jackson, G. and Vargha-Khadem, F. (1992) Possible blindsight in infants lacking one cerebral hemisphere. *Nature*, **360**, 461–463.
7.   Cowey, A. and Stoerig, P. (1995) Blindsight in monkeys. *Nature*, **373**, 247–249.
8.   Whittington, M.A., Traub, R.D. and Jefferys, J.G. (1995) Synchronized oscillations in interneuron networks driven by metabatropic glutamate receptor activation. *Nature*, **373**, 612–615.
9.   Castiello, U., Scarpa, M. and Bennett, K. (1995) A brain-damaged patient with an unusual perceptuomotor deficit. *Nature*, **374**, 805–808.
10. Barbur, J.L., Watson, J.D.G. and Frackowiak, R.S.J. (1993) Conscious visual perception without V1. *Brain*, **116**, 1293–1302.
11. Stoerig, P. (1996) Varieties of vision: from blind responses to conscious recognition. *Trends in the Neurosciences*, **19**, 401–406.
12. Franken, P., Dijk, D.J., Tobler, I. and Borbely, A.A. (1994) High-frequency components of the rat electrocorticogram are modulated by the vigilance states. *Neuroscience Letters*, **167**, 89–92.
13. Pinault, D. and Deschenes, M. (1992) Control of 40 Hz firing of reticular thalamic cells by neurotransmitters. *Neuroscience*, **51**, 259–268.
14. Hopfield, J.J. (1995) Pattern recognition computation using action potential timing for stimulus representation. *Nature*, **376**, 33–36.
15. de Charms, R.C. and Merzenich, M.M. (1996) Primary cortical representation of sounds by the coordination of action-potential firing. *Nature*, **381**, 610–613.
16. Heiss, W.D., Weinhard, K., Graf, R., Lottgen, J., Pietrzyk, U. and Wagner, R. (1995) High-resolution PET in cats: application of a clinical camera to experimental studies. *Journal of Nuclear Medicine*, **36**, 493–498.
17. Magata, Y., Saji, H., Choi, S.R., Tajima, K., Takagaki, T., Sasayama, S., Yonekura, Y., Kitano, H., Watanabe, M. and Okada, H. (1995) Noninvasive measurements of cerebral blood flow and glucose metabolic rate in the rat with high-resolution animal positron emission tomography (PET): a novel in vivo approach for assessing drug action in the brains of small animals. *Biological and Pharmaceutical Bulletin*, **18**, 753–756.
18. Torres, E.M., Fricker, R.A., Hume, S.P., Myers, R., Opacka-Juffry, J., Ashworth, S., Brooks, D.J. and Dunnet, S.B. (1995) Assessment of striatal graft viability in the rat in vivo using a small diameter PET scanner. *Neuroreport*, **6**, 2017–2021.

19. Bruehl, C., Kloiber, O., Hossman, K.A., Dorn, T. and Witte, O.W. (1995) Regional hypometabolism in an acute model of focal epileptic activity in the rat. *European Journal of Neuroscience*, **7**, 192–197.

20. Nagel, T. (1974) What is it like to be a bat? Reprinted in Hofstadter, D.R. and Dennett, D.C. (1981) *The Mind's I*, pp. 391–403. London: Penguin Books.

21. Churchland, P.S. (1986) *Neurophilosophy*. Cambridge: Bradford Books, M.I.T. Press.

22. Ebbesson, S.O.E. (1980) *Comparative Neurology of the Telencephalon*. New York: Plenum Press.

23. Llinas, R. and Precht, W. (1976) *Frog Neurobiology*. Berlin: Springer-Verlag.

24. Ventriglia, F. (1994) *Neural Modeling and Neural Networks*. New York: Pergamon Press.

25. Damasio, A.R. (1994) *Descarte's Error*. New York: Putnam Books.

CHAPTER 7   CONSCIOUSNESS AND QUANTUM
MECHANICS

Quantum mechanics provides a theory for the behavior of elementary particles and atoms. Even though it has been enormously successful in this task, the conceptual foundations of the theory are still a matter of hot debate. Mysterious dilemmas emerge as one probes deeper into the meaning of quantum mechanical ideas. Given that consciousness is itself such a mystery, some pundits have suggested that the spate of publications over the last decade on the possibility that quantum mechanical principles are needed to explain consciousness arise from the conviction that two such fundamental mysteries must be related! Perhaps the foremost exponent of the idea that quantum mechanical effects are involved in consciousness is the mathematical physicist Roger Penrose[1]. However, Francis Crick[2] dismisses the idea entirely. This chapter sets out the arguments so far developed for a role of quantum mechanical effects in consciousness[3,4]. No quantum principles have yet been needed to explain any neuroscientific phenomenon observed in the laboratory, with the possible exception of the interaction of photons with the photoreceptors of the retina. But then, is consciousness a laboratory phenomenon?

7.1 Quantum mechanics

The problem which confronted the founding fathers of quantum mechanics (Figure 7.1), concerning the behavior of subatomic particles, is best illustrated by means of the following experiment. Consider an apparatus consisting of a heating element that acts as a source of electrons which are ejected first through a hole A1, and subsequently through a very narrow hole A2, as shown in Figure 7.2. The dots in this figure indicate the pattern built up on a scintillation screen of the positions of electron collisions when many pulses of the electron gun have been activated. The extent of the scatter becomes greater as the hole A2

144

Figure 7.1 The founders of quantum mechanics in discussion together, 1962. Niels Bohr (left), Werner Heisenberg (middle) and Paul Dirac.

becomes smaller than a certain value. An additional hole A3 is now added to the apparatus, as shown in Figure 7.2B. In this case the expected pattern of electron collisions is simply the addition of two patterns like that in Figure 7.2A. However, the actual observed pattern of collisions is quite different, consisting of regions of high density collisions alternating with regions in which there are no collisions at all.

This surprising result has an analogy with the propagation of waves through holes. Figure 7.2Ca shows wave fronts (vertical lines) propagating through a hole which is much larger than their wavelength, whereas Figure 7.2Cb shows that the waves spread out when they propagate through a hole that is of similar or smaller size than their wave length; this is analogous to the increased scattering of electron collisions as the hole becomes smaller in Figure 7.2A. In Figure 7.2Cc the waves are made to propagate through two holes; in this case the waves through one hole interfere with the waves through the other, so that in some cases they reinforce each other and in other cases negate each other. The consequence, as shown, is that there are regions where the waves are strong and these alternate with regions in which there are no waves at all. The situation is analogous to that of the pattern of electron collisions shown in Figure 7.2B.

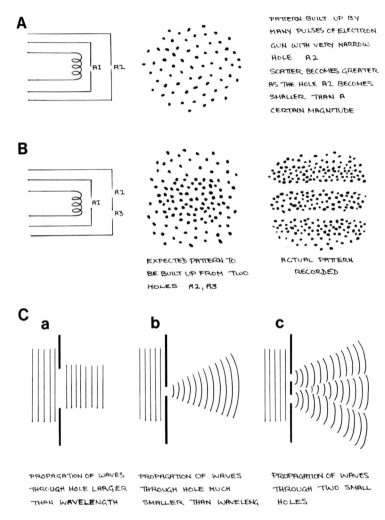

A PATTERN BUILT UP BY MANY PULSES OF ELECTRON GUN WITH VERY NARROW HOLE A2 SCATTER BECOMES GREATER AS THE HOLE A2 BECOMES SMALLER THAN A CERTAIN MAGNITUDE

B EXPECTED PATTERN TO BE BUILT UP FROM TWO HOLES A2, A3   ACTUAL PATTERN RECORDED

C a PROPAGATION OF WAVES THROUGH HOLE LARGER THAN WAVELENGTH

b PROPAGATION OF WAVES THROUGH HOLE MUCH SMALLER THAN WAVELENG

c PROPAGATION OF WAVES THROUGH TWO SMALL HOLES

Figure 7.2   The quantum mechanical behaviour of subatomic particles.

7.2 Bohr's interpretation of Schrodinger's wave equation

Schrodinger's wave equation provides a quantum mechanical description of this behavior of electrons and is interpreted in the following way by the adherents of Niels Bohr of Denmark (Figure 7.3); it is for this reason called the Copenhagen interpretation[5,6]. An electron is considered to be in a superposition

Figure 7.3   Niels Bohr (1885–1962).

of the different states of the observable which is to be measured; for example, its collision position on the scintillation screen in Figure 7.2B. That is, the electron is considered to be at several different positions on the screen, as given by the solutions of Schrodinger's equation. The measuring apparatus (namely the scintillation screen) used to determine the position is also in a state of superposition with the different states of the observable, namely the collision positions. According to this interpretation, when an observer looks at the scintillation screen a superposition of observable states occurs involving the screen and the observer's brain. This piling on of states of superposition can only be terminated at the level of consciousness. All these correlated states can be ascertained by the deterministic Schrodinger wave equation. The question then arises as to where does the discontinuity occur which leads to the collapse of all the states except that one associated with a particular position on the scintillation screen, namely that observed? Schrodinger's wave equation for the electrons passing through the two holes of the partition in the experiment of Figure 7.2B therefore has solutions which give non-zero values for the position of electron collisions in those areas in which a position measurement might ultimately find a collision has occurred. In the standard Copenhagen dogma, the superposed positions of the electron collision all exist in reality up to the moment that a measuring device is used by an observer to determine the collision site; at this time the wave function collapses to zero at all sites except one[7].

The extraordinary implications of the Copenhagen interpretation of quantum mechanics is illustrated by means of the famous thought experiment named Schrodinger's cat, which is illustrated in Figure 7.4. Electrons are fired at a partition with two holes as in Figure 7.2B. If the electrons appear in the middle of the scintillation screen, they trigger via a light sensing photoelectric tube the release of a hammer which smashes a glass capsule of cyanide that kills a cat which is resident in the same closed box as the apparatus. The position of the electron on the scintillation screen is a quantum mechanical problem, so the solution of Schrodinger's equation gives the superposition of states of the electron being at the different positions on the screen. The photoelectric tube and the hammer also have correlated states with the electron positions that are given by Schrodinger's equation; these involve whether the tube is excited to give a current or not, for instance, and the position of the hammer, for instance. Finally, the cat is part of this quantum system with states in which it is dead or alive. In this isolated quantum system, the paradox then arises that the cat is neither dead nor alive. The Copenhagen interpretation is that this chain, called the von Neumann chain, is completed by an observer who looks into the isolated room. The conscious observer then becomes part of the quantum system. According to Eugene Wigner[8], this chain is broken at the level of the mind, so that all the different states given by Schrodinger's equation collapse but one,

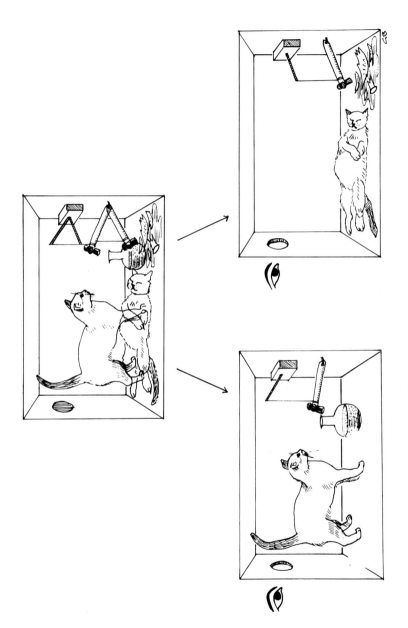

Figure 7.4   Schrodinger's cat highlights the dilemmas presented by the Copenhagen interpretation of quantum mechanics.

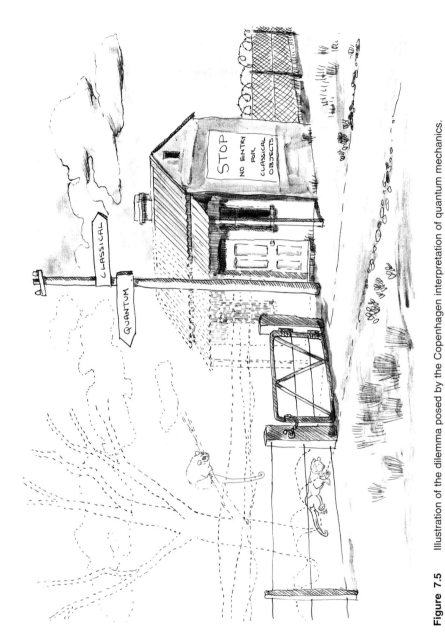

Figure 7.5   Illustration of the dilemma posed by the Copenhagen interpretation of quantum mechanics.

and the cat is found either dead or alive. In practice, the Copenhagen interpretation of quantum mechanics draws an arbitrary line of demarcation between the quantum world (of say atomic particles) and the classical world (of say measuring instruments and humón brains), as illustrated by the cartoon of Figure 7.5. Application of Schrodinger's equation is then confined to the quantum world.

7.3 Bohm's interpretation of Schrodinger's wave equation

David Bohm's[9, 10] interpretation of Schrodinger's equation escapes these dilemmas of the Copenhagen school. This is illustrated by the following experiment given in Figure 7.6. A laser sends a photon to a beam splitter at which the photon's wave function splits and passes along pathways A and B, whereas the photon continues only along path A. The photon is then reflected with its wave function into a switching box. If the switch is on, then the part of the wave function which took the same route as the photon, together with the part which did not, meet at the box where the arrows converge and give the interference patterns shown on the screen; these patterns indicate the probability distribution of where the photons impact on the screen. If the switch is off, then the part of the wave function that did not take the same route as the photon never meets the photon again; this is decisive information for the photon, which then acts as a particle and is recorded as such by the photon-detector. In this interpretation of Schrodinger's equation there are no superimposed states that lead to the difficulties which arise in the Copenhagen school's ideas[11].

Schrodinger's cat escapes the dilemma of being neither dead nor alive in David Bohm's intepretation of how the Schrodinger wave equation should be applied in quantum mechanics. In this case, the position of the electron on the

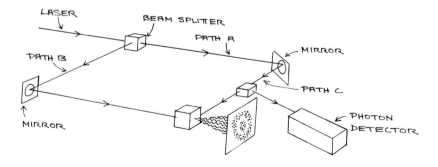

Figure 7.6   Bohm's interpretation of Schrodinger's equation.

scintillation screen, after it has passed through one of the two double slits, is not given by the superposition of states of the electron at different positions on the screen, so that the electron is not simultaneously at different sites before an observation is made. Rather, the position of the electron on the screen is ascertained deterministically in the following way. The Schrodinger wave function of the electron splits up and passes through both slits in the screen, while the electron itself passes through one of the slits as a single particle must do. Which slit the electron passes through is deterministic, based simply on the initial position and initial wave function, according to the linear differential equations of motion. The two parts of the electron's wave function join on the other side of the double-slit screen, with the part of the wave function that did not take the same pathway as the electron itself interfering with the wave function that went with the electron. It is the interference of these two waves which then determines the position at which the electron hits the screen (the part of the wave function that did not go with the electron is not capable of conveying effects to any other particle). This deterministic procedure does not involve the metaphysical paradoxes which arise in the case of the Copenhagen interpretation with its superposition of different states. According to Bohm's theory, the electron has either hit the middle of the screen or not, so that the cat is dead or alive before the observer looks in the box. The cat is not in a superposition of states in which it is neither dead nor alive.

7.4 Gell-Mann and Zurek's interpretation of Schrodinger's wave equation

Schrodinger's cat also escapes the dilemma of being neither dead nor alive in the Gell-Mann[12] and Zurek[13] interpretation of how Schrodinger's equation should be interpreted. In this case it is argued that the quantum-system residing in the Schrodinger cat cage is not a closed system, isolated from the environment even in the absence of an observer. Indeed, no macroscopic system is isolated from its environment. This then avoids the difficulty that the deterministic linear Schrodinger wave equation of a system evolves into a state that simultaneously contains many possibilities which are never seen to coexist, such as the cat being both dead and alive. The core problem is the principle of superposition that resides in Schrodinger's wave equation. Gell-Mann and Zurek argue that while Wigner is correct in saying that the final point in the von Neumann chain resides in our consciousness of the world, that is we are conscious of only one of the superposition states given by the Schrodinger equation, it is very likely that the state is already chosen from amongst the superpositions before consciousness is involved at all. In the cases of Schrodinger's cat, the superpositions are destroyed by interaction between the quantum events in the cage and the environment, well before an observer looks in the cage. It is argued that the environment effectively makes a superselection which prevents certain superpositions from occurring, so that only states that survive this process become classical and so are observed. This is the reason why just one of the quantum alternatives is perceived. In a sense, the interaction of the quantum system with the environment involves a transfer of information from the quantum system to the environment. If an observer is present, then such leakage of information from the quantum system will be to the brain of the observer, in this case constituting part of the environment. Indeed, quantum events in the brain, say involving different quantum superpositions of the states of a neuron, will be lost very quickly compared with the times of biological processes. This is due to information leaking from the neuron to the environment (in this case consisting of, saû, ions surrounding the neuron), leaving the neuron in a preferred classical state.

According to this scheme, observers have lost their privileged position in monitoring a system involving the transfer of information to a recording apparatus. Information is also transferred from the quantum system to other elements of the environment. The only difference between the two is the difficulty of decoding the information taken up by elements of the environment compared with that taken up by the observer's specially designed observing apparatus. In the case of the two slit experiment, by the time the electron reaches the screen it will, under normal circumstances, have interacted with components of the environment that greatly decrease the probability that superimposed states will exist at the screen. Only definite states will exist; that is, the electron will have hit the screen within one of the three regions shown in Figure 7.2B.

It is usually argued that the brain is too hot at body temperature to maintain the coherent states of superposition for a neuron which might arise from interference effects at the quantum level. Gell-Man and Zurek contend that even if such interference effects exist, giving rise to coherent states of superposition, they will very quickly (compared with biological times) dissipate due to interactions between the neuron and its environment, so removing all but one of the solutions to the Schrodinger equation.

7.5 The concept of non-local reality in quantum mechanics

All interpretations of quantum mechanics so far enumerated require the concept of non-local reality[14,15]. In non-local reality an event that occurs in, say, region A can instantaneously have a physical effect in region B, independent of how far A is from B and of the conditions in the space between A and B. This is illustrated in Figure 7.7 by the following simple experiment. A source of photons is placed midway between two polarizing films which allow the passage of all photons that are linearly polarized along a horizontal axis. However, they allow

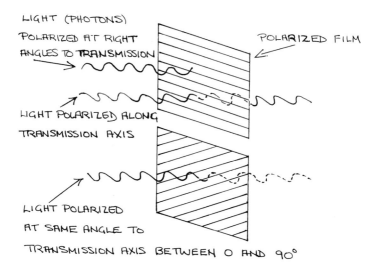

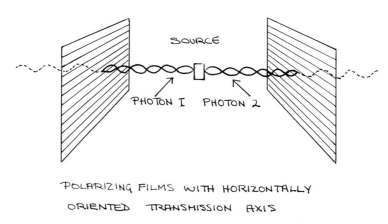

Figure 7.7   The concept of non-local reality.

with decreasing probability the passage of photons that are linearly polarized at an angle q to the horizontal (where the probability is cos²q). The source contains photons that each have a superposition of states in which they are linearly polarized along a horizontal axis and a vertical axis; in this case q is 45° so that cos²q = 0.5 and there is a 50% probability that a photon will pass through the horizontally oriented polarizing film. If pairs of such correlated photons are simultaneously shot out of the source at both screens, then the expectation is that each should independently pass through with a 50% probability. Instead, what happens is that if one photon is transmitted so is the other and if one is blocked so is the other. The second photon of the pair 'knows' what the other photon does even though they are well separated. There is instantaneous transmission of information from one photon to the other. The discovery that such correlated particles can affect each other over large distances independent of intervening space amounts to a non-local interaction and is therefore termed 'non-local reality'.

It has been suggested by Penrose that an example of non-local reality at work can be observed in the assembly of quasi crystals[16]. For instance, the Al-Li-Cu alloy has a crystal symmetry that apparently can only be assembled by using non-local ordering of the atoms[17,18]. In this case, to assemble such a crystal requires that the pattern of atoms some distance from the region of assembly be determined in order to avoid errors in the arrangement of atoms gathered up at the point of assembly. This could be done by a mechanism of non-local reality, as described in relation to Figure 7.7.

7.6 Ideas concerning the operation of non-local reality in the brain: the origins of consciousness?

Eccles[19,20] believes that nerve terminals may be considered to have a crystal-like structure, so that non-local phenomena occur by quantum mechanical means at these terminals. The physical structure of nerve terminals and of the site where they attach to the surface of neurons was first given emphasis by Sherrington[21] (Figure 7.8), who first coined the word 'synapse'. Although the phrase 'synaptic connection' has been used repeatedly in previous chapters, there has not been much need to consider the synapse itself in any detail, with the exception of synaptic receptors described in relation to schizophrenia in Chapter 5. Synapses form all over the surface of neurons, as is illustrated in Figure 7.9 by considering a neuron from the CA3 region of the hippocampus. This neuron has synapses (S) on its round cell body as well as on the large dendritic shafts; the largest synapse in the brain, formed by the so-called mossy fibre synapses (MF), are also illustrated. Members of the class of large neurons in the brain are called pyramidal cells, and each may possess up to 30,000 synapses on its surface. Collections of small vesicles are crowded together in each bouton as was illustrated in Figure 5.10. As described in Chapter 5, small vesicles contain transmitter substances which, if released from the bouton, diffuse onto the membrane of the neuron and transiently bind to it, therebye increasing or decreasing the electrical excitability of the neuron.

A synaptic bouton may then have a paracrystalline structure which requires Schrodinger's equation to describe its quantum states. Shown in Figure 7.10 is a

Figure 7.8   C.S. Sherrington (1858–1952).

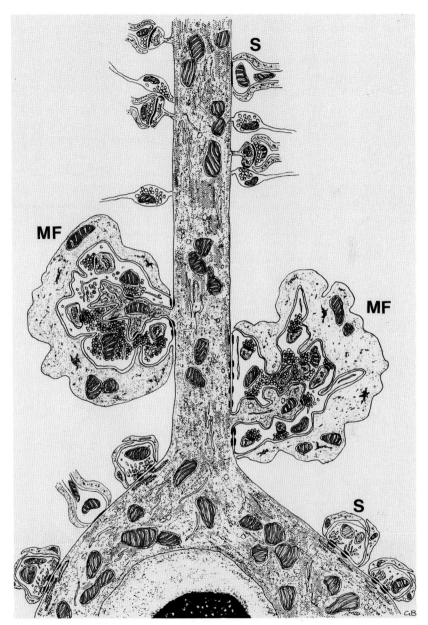

Figure 7.9 A neuron from the CA3 region of the hippocampus together with synapses on its round cell body as well as on the large dendritic shafts.

cut away indicating the contents of a nerve terminal forming a synapse. The membrane of the synaptic bouton immediately impinging on a dendrite possesses protrusions called dense projections (dp) which constitute what is called the active zone of the bouton. Eccles suggests that these dense projections make up a paracystalline structure. Large numbers of synaptic vesicles (SV) insert between the projections of the active zone. It is hypothesized that following arrival of a nerve impulse in the bouton, a single vesicle on the active zone may release its contents by the method of exocytosis. Other vesicles (about 50 to 100 on the zone) are prevented from releasing because the hypothesized paracystalline array of the active zone possesses components that are in a state of quantal coherence, enabling non-local instantaneous interactions of different parts of the active zone. Once a vesicle, interacting with the active zone, is irretrievably committed to the release of its transmitter, then this information is conveyed instantaneously to the rest of the active zone, producing non-local changes in the paracystalline array that prevent other vesicles from releasing transmitter.

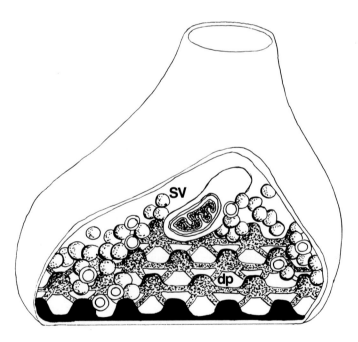

Figure 7.10   The contents of a nerve terminal.

Non-local reality may arise in a different way in the brain through quantum correlations involving phonons, as first suggested by Frohlich[22]. Phonons stand in relation to the electrical vibration of matter (say a dipole molecule) in an analogous way to photons and oscillations of the electromagnetic field. Phonons can be thought of as particles then in the same way as photons, and so are subject to quantum mechanical effects in the same way as photons. How may phonons be used by neurons? One possibility involves consideration of the way in which the proteins in the membrane of neurons, which are responsible for the movement of ions between the inside and outside of the neuron, interact with the intracellular skeleton of the neuron. The components of these membrane proteins may be considered to be dipole molecules that give rise to phonons. If the rate of energy supply to these oscillating dipoles from the cytoskeleton is sufficiently high, a fraction of the energy is not lost to the elements that make up the constant temperature heat bath surrounding these membrane proteins, such as ions moving in the solutions on either side of the neuron's membrane. This energy fraction may then be stored in such a way as to create phase and amplitude correlations between the phonons of the system over very large distances. In other words, the energy fraction is stored in the phonons of the system and may be used to create new phonons[23]. Quantum correlations arise between the phonons in this way so that coherent superpositions occur in the system of phonons over large distances. Frohlich[24] and Marshall[25] have argued that quantum correlations between phonons of the neuron's membrane channel proteins responsible for the excitability of the neurons could arise over large distances in the nervous system.

Non-local reality, however it arises, has been suggested by Penrose[1] to mediate the holistic experiences of consciousness that involve correlations between near simultaneous events which are large distances apart in the neocortex. Such events may involve identification of an object in the inferior temporal cortex with determination of its movement in the parietal cortex, many centimeters away. These quantum correlations over extensive distances may then be responsible for the coherence of conscious experience involving so many of the brain's parallel processing neuronal units which are spread throughout the neocortex.

The possibility that quantum mechanical ideas, which are so removed from 'commonsense', will be required to explain some aspect of synaptic or neuronal function is still open. Although at present this is not the case, we are still at such a profound level of ignorance concerning synaptic, neuronal and brain function that it would be very premature to say the case is closed for quantum mechanical explanations of consciousness. The high temperature of the brain would seem to be a large obstacle for the emergence of any states of quantum coherence. However, the recent success in obtaining superconductivity at relatively high temperatures, involving as it does quantum mechanical effects, does suggest that even the temperature problem is probably not an insurmountable obstacle. It may well be that the weird world of the quantum does include our consciousness.

7.7 References

1.   Penrose, R. (1991) *The Emperor's New Mind*. Oxford: Oxford Press.
2.   Crick, F. (1994) *The Astonishing Hypothesis*. London: Simon & Schuster.
3.   Hodgson, D. (1991) *The Mind Matters*. Oxford: Clarendon Press.
4.   Lockwood, M. (1989) *Mind, Brain and Quantum*. Oxford: Basil Blackwell.
5.   Pais, I. (1991) *Niels Bohr's Times*. Oxford: Oxford University Press.
6.   Von Neumann, J. (1955) *Mathematical Foundations of Quantum Mechanics*, Princeton N.J.: Princeton University Press.
7.   Sewell, G.L. (1986) *Quantum Theory of Collective Phenomenon*. Oxford: Oxford University Press.
8.   Wigner, E.P. (1962) In *The Scientist Speculates: An Anthology of Partly-Baked Ideas*, edited by I.J. Good, pp. 284-302. London: Heinemann.
9.   Bohm, D. (1957) *Causality and Chance in Modern Physics*. London: Routledge & Kegan Paul.
10. Bohm, D. and Hiley, B.J. (1993) *The Undivided Universe*. London: Routledge.
11. Albert, D.Z. (1994) Bohm's alternative to quantum mechanics. *Scientific American*, May, 32–39.
12. Gell-Mann, M. & Hartle, J.B. (1989) Quantum mechanics in the light of quantum cosmology. In *Proceedings of the 3rd International Symposium on the Foundation of Quantum Mechanics*, edited by S. Kobyashi. Physical Society of Japan, Tokyo.
13. Zurek, W.H. (1991) Decoherence and the transition from quantum to classical. *Physics Today*, October, 36–44.
14. Aspect, A., Dalibard, J. & Gerard, R. (1982) Experimental test of Bell's inequalities using time-varying analyzers. *Physical Reviews Letters*, **49**, 1804–1807.
15. Bell, J.S. (1986) Six possible worlds of quantum mechanics. Proceedings of the Nobel Symposium 65; Possible Worlds in Arts and Sciences, Stockholm, August 11–15.
16. Gardner, M. (1989) *Penrose Tiles to Trap Door Ciphers*. New York: W.H. Freeman & Company.
17. Penrose, R. (1989) Tilings and quasi crystals. In *Aperiodicity and Order 2*, edited by M. Jaric. London: Academic Press.
18. Shechtman, D., Blech, I., Gratias, D. & Cahn, J.W. (1984) Metallic phase with long-range orientational order and no translational symmetry. *Physical Review Letters*, **53**, 1951–1953.
19. Beck, F. & Eccles, J.C. (1992) Quantum aspects of brain activity and the role of consciousness. *Proceedings of the National Academy of Science of USA*, **89**, 11357–11361.
20. Eccles, J.C. (1993) *How the Self Controls its Brain*. Berlin: Springer Verlag.
21. Granit, R. (1966) *Charles Scott Sherrington: An Appraisal*. London: Tomas Nelson & Sons Ltd.
22. Frohlich, H. (1968) Long-range coherence in biological systems. *International Journal of Quantum Chemistry*, **2**, 641–649.
23. Bond, J.D. and Huth, G.C. (1986) Electrostatic modulation of electromagnetically

induced nonthermal responses in biological mechanisms. In *Modern Bioelectrochemistry*, edited by F. Gutmann and H. Keyze, pp. 283–313. New York: Plenum.

24. Frohlich, H. (1986) Coherent excitation in active biological systems. In *Modern Bioelectrochemistry*, edited by F. Gutman & H. Keyzer, pp. 241–261. New York: Plenum.

25. Marshall, I.M. (1989) Consciousness and Bose-Einstein condensates. *New Ideas in Psychology*, **7**, 73–83.

AN EPILOGUE: BLADE RUNNER OR EINSTEIN?

What is a neuroscientist doing writing a book on consciousness, a topic that is often taken as exclusively within the scope of philosophers? The nervous system is the domain of the neuroscientist, and as the brain is a neural mechanism that is related to consciousness in a most intimate way, the subject falls within the purview of this scientific discipline. A number of questions concerned with the phenomenon of consciousness have therefore been considered from a neuroscience point of view. First, the general problem of how the brain might cope with syntax and semantics has been examined together with the critical question of whether it is possible that qualia could ever be understood by neuroscience. Next, an attempt was made to identify the neural processes that are the concomitants of the visual experience of attending to or being aware of an object, as well as the process of how we have sensations in consciousness of movement and of exerting forces with our limbs. The possible origins of various distortions of consciousness, such as visual hallucinations that arise in psychotic states like schizophrenia, were then detailed. This then led to a study of the possibility that species other than our own possess consciousness, and what form this might take. Finally, an analysis was presented of whether quantum mechanical principles might be involved in the workings of the synaptic connections which form the neural networks of the brain, with the expectation that if they are, then this might offer some insight into the problem of how qualia arise.

A neuroscience of consciousness is now possible. Techniques involving the psychophysical studies of *Homo sapiens* with neocortical lesions, the measurement of 40 Hz oscillations in the magneto-encephalogram, and imaging of the neocortex with PET and fMRI offer the prospect of studying the likely sites of consciousness in *Homo sapiens* as well as in other species. There appear to be two main objections to this approach. The first is that posed in the Preface,

162

namely the philosophical problem of whether consciousness can be considered to be the biological workings of the brain. This is really not central to a neuroscience of consciousness which seeks to ascertain the sites of conscious experience in the brain of *Homo sapiens* and in that of other species, should they possess consciousness. Such a neuroscience does not attempt to answer the question of whether consciousness *is* the workings of some particular part of the brain, although it may well turn out to illuminate the problem in unforseen ways.

The other difficulty with a neuroscience of consciousness is that it is not possible to argue about consciousness at all in species that do not have language, that is all species other than *Homo sapiens*, as there can be no direct report of the contents of consciousness. The studies on monkeys enumerated in Chapter 6 suggest that this need not be an impediment. Consciousness is obviously assumed present amongst the deaf and dumb. This is inferred by analogy with the fact that they belong to a species that does possesses consciousness. However, the conferring of consciousness on other members of our species becomes somewhat tenuous when it is considered that in some cases the contents of their consciousness is very restricted indeed. Such is the case with the patient, described in Chapter 6, who was unable to see more that one object at the same time from a given semantic class and who could only manipulate semantically different objects. It may also be very restricted in infants with or without brain lesions. There is a level of inference involved in conferring consciousness on deaf mutes. This kind of inference is also necessary when conferring consciousness on species other than *Homo sapiens*. Such kinds of inferences occur in arguments about supporting evidence for the evolution of species: we cannot rerun the evolutionary scenario as it has most likely occurred, but infer from the fossil record what seems to be support for a very reasonable hypothesis. A neuroscience of consciousness would do the same: arrive at reasonable conclusions concerning the existence of consciousness and its phenomenal content in the neural networks that comprise the telencephalon, diencephalon and mesencephalon of different species.

Why bother with a neuroscience of consciousness? Besides the intrinsic fascination with bringing together observations associated with consciousness`in *Homo sapiens* and other species, a comparative discipline of this kind would reveal a great deal about the workings of our own brains. It would emphasize in the most dramatic way that the consciousness which we have taken for granted as unique to *Homo sapiens* is a complex with many different levels of sophistication, generated in different parts of our brains and also in the brains of other species. The naming of *Homo sapiens*, that is wise man, has conferred on the species something which might be a misnomer, namely that it alone possesses the wisdom associated with consciousness. A neuroscience of consciousness might show that this phenomenon has evolved. We do not then have to seek for consciousness in another species on other planets for *Homo sapiens* may have been surrounded by it all the time.

According to the philosophers Searle and Chalmers, the problem considered in this book, of examining the extent to which neuroscience has explained consciousness, is ill-posed. All that neuroscience can do is explain the neural correlates of consciousness, leaving the difficult problem of explaining what consciousness *is* untouched. In defence, it should be mentioned that there is a strong metaphysical tradition in western philosophy, going back to Plato, of seeking for a set of principles that will provide a unifying explanation of the physical world and consciousness. What if the dualistic notion is correct, and consciousness is not to be identified with the synaptic workings of the brain? In that case, research will be needed to discover the grand laws that underpin both phenomena. This is a tall order, given the genius that has been required so far to establish the laws of the natural world. It will require talent at least as great as that of Einstein to bring forth explanations from a well of originality that is without precedent, as was the case with his general theory of relativity. Indeed, it raises the question of whether the synaptic networks of the brain of *Homo sapiens* have evolved to a level capable of providing a solution to such a problem. For neuroscientists there is still the realization that the synaptic networks of the brain work in a parallel relationship with that of consciousness, so that the scientific study of the former gives insights into the contents of the latter even though it might not provide an explanation of how consciousness arises.

Attempts to produce a robot that can use a reasonably sophisticated seeing device have not yet been successful. The problem is a deep one, as *Vision* by David Marr made abundantly clear many years ago. He pointed out the extent of understanding of some of the synaptic network processes subserving vision which is required before the possibility of building a seeing robot might be realized. However, neuroscience is approaching this goal, so that a seeing robot seems feasible. This raises the general question as to whether the duplication of parts of the brain in silicon rather than neurons is likely to be able to carry out the same functions as a biological brain. This certainly seems possible for the early stages of sensory processing, such as that of the retina, thalamus and perhaps parts of the occipital cortex in the case of vision. It might, however, be extremely difficult for the higher areas of brain functioning concerned with, for example, semantics as discussed in Chapter 2. Phillip Dick describes in *Do Androids Dream of Electric Sheep*, made into the film *Blade Runner*, a situation in which the androids cannot in general be distinguished from *Homo sapiens*. If a situation like this should ever arise, then Turing would argue that consciousness has been created. However, even though knowledge of synaptic networks might eventually become such as to enable the construction of androids, that would not necessarily mean that an understanding of the origins of consciousness had been reached, as David Chalmers has emphasized. Nevertheless, in the absence of insights provided by a theory that encompasses both the physical world and consciousness, *Homo sapiens* will almost certainly accept the androids as possessing consciousness like ourselves if they pass all the tests that can be applied to confirm if a species possesses this attribute. If this should ever eventuate, we will have been seduced into accepting consciousness as an attribute of certain kinds of synaptic networks, without perhaps ever understanding why that should be so.

The first purpose of this book has been to present the contributions which neuroscience has recently made to our understanding of where the different contents of consciousness arise in the brain. The second has been to see how the workings of the synaptic connections in the brain might give rise to these contents. Arguments have been put forward that the concepts and techniques which neuroscience now has available offer the opportunity to identify and explore those parts of the brain which generate the different contents of consciousness. This might be possible not only for our own species, but for others as well. Such a neuroscience of consciousness will provide insights into the origins of this phenomenon. Perhaps more importantly, it will give us the means of ameliorating those distortions of consciousness which arise in psychotic conditions that are responsible for so much mental suffering in the world.

SOURCE OF ILLUSTRATIONS

Chapter 3   **The Holistic Nature of Consciousness**

Figure 3.3   Reprinted from *Proceedings of the National Academy of Science of USA*, Gray, A.K. & Singer, W., 'Stimulus-specific neuronal oscillations', **86**, Figure 1A, p. 1699, Copyright 1989, with kind permission of Professor W. Singer.

Figure 3.5   Data in the photomicrograph from Siegrid Lowel.
Electrophysiological data reprinted from *IBRO News*, Singer, W., 'Response synchronization of cortical neurones', **19(1)**, Figure 2, p. 7, Copyright 1991, with kind permission from Elsevier Science Ltd, The Boulevard, Langford Lane, Kidlington OX5 1GB, UK.

Figure 3.6   Reprinted with permssion from Engel, A.K., Konig, P., Kreiter, A.K. and Singer, W., Interhemispheric synchronization of oscillatory neuronal responses in cat visual cortex. *Science*, **252**, Figure 3, p. 1178. Copyright 1991 American Association for the Advancement of Science.

Figure 3.7   Data from Konig, P. and Schillen, T.B. 'Stimulus-dependent assembly formation of oscillatory responses: 1. Synchronization', *Neural Computation*, **3:2** (Summer, 1991), pp. 155–156. © 1991 by the Massachusetts Institute of Technology.

Figure 3.8   Figure adapted from Konig, P. and Schillen, T.B. 'Stimulus-dependent assembly formation of oscillatory responses: 1. Synchronization', *Neural Computation*, **3:2** (Summer, 1991), pp. 155–156. © 1991 by the Massachusetts Institute of Technology.

Figure 3.9   Reprinted with permission from Konig, P. and Schillen, T.B. 'Stimulus-dependent assembly formation of oscillatory

Chapter 4   The Consciousness of Muscular Effort and Movement

Chapter 5   The Distortion of Consciousness

Figure 5.1   Reprinted with permission from Churchland, P.S. & Sejnowski, T.J., Perspectives on cognitive neuroscience, *Science*, **242**, Figure 1, p. 742. Copyright 1988 American Association for the Advancement of Science.

Figure 5.2   Reprinted with permission from *Nature*. Tootell et. al., 'Visual motion after effect in human cortical area MT revealed by functional magnetic resonance imaging', **375**, Figures 1 and 2, p. 140 and Figure 3, p. 141. Copyright (1995) Macmillan Magazines Limited.

Figure 5.4   Reprinted with permission from *Nature*. Silbersweig et. al., 'A functional neuroanatomy of hallucinations in schizophrenia', **378**, Figure 1, p. 177. Copyright (1995) Macmillan Magazines Limited.

Figure 5.5   Reprinted with permission from *Nature*. Silbersweig et. al., 'A functional neuroanatomy of hallucinations in schizophrenia', **378**, Figure 2, p. 178. Copyright (1995) Macmillan Magazines Limited.

Figure 5.7   Reprinted from *Psychopharmacology*, Joyce, J.N., 'The dopamine hypothesis of schizophrenia: limbic interactions with serotonin and norepinephrine', **112**, Figure 4, p. 523, (1993), with kind permission of Elsevier Science - NL, Sara Burgerhartstraat 25, 1055 KV Amsterdam, The Netherlands.

Figure 5.8   Reprinted from *Psychopharmacology*, Joyce, J.N., 'The dopamine hypothesis of schizophrenia: limbic interactions with serotonin and norepinephrine', **112**, Figure 5, p. 526, (1993), with kind permission of Elsevier Science - NL, Sara Burgerhartstraat 25, 1055 KV Amsterdam, The Netherlands.

Figure 5.10   Reprinted with permission from Farde, L., 'The advantage of using positron emission tomography in drug research', *Trends in Neuroscience*, **19**, Figure 4, p. 214, 1996.

Chapter 6   The Evolution of Consciousness

Figure 6.1   Dennett, D.C. (1996) *Kinds of Minds*. New York: Basic Books. Harper Collins.

Figure 6.2   Romer, A.S. (1960) *The Vertebrate Body*. London: W.B. Saunders and Co.

Figure 6.3   After Figure 366 in Romer, A.S. (1960) *The Vertebrate Body*. London: W.B. Saunders and Co.

Figure 6.4   After Figure 365 in Romer, A.S. (1960) *The Vertebrate Body*. London: W.B. Saunders and Co.

Figure 6.5   After Figure 383 in Romer, A.S. (1960) *The Vertebrate Body*. London: W.B. Saunders and Co.

Figure 6.6   After Figure 13.1 in Sarnat, H.B. and Netsky, M.G. (1981) *Evolution of the Nervous System*. New York: Oxford University Press.

Figure 6.7   After Figure 13.5 in Sarnat, H.B. and Netsky, M.G. (1981) *Evolution of the Nervous System*. New York: Oxford University Press.

Figure 6.8   From Figure 2 in Weiskrantz, L., Barbur, J.L. and Sahraie, A. (1995) Parameters affecting conscious versus unconscious visual discrimination with damage to the visual cortex (V1). *Proceedings of the National Academy of Science, USA*, **92**, 6122–6126.

Figure 6.9   Reprinted with permission from *Nature*. Braddick, Braddick, O., Atkinson,J., Hood, B., Harkness, W., Jackson, G. and Vargha-Khadem, F., 'Possible blindsight in infants lacking one cerebral hemisphere', **360**, Figure 1, p. 462. Copyright (1992) Macmillan Magazines Limited.

Figure 6.10   Reprinted with permission from *Nature*. Cowey & Stoerig, 'Blindsight in monkeys', **373**, Figures 2 and 3, p. 248. Copyright (1995) Macmillan Magazines Limited.

Figure 6.11   Adapted from Figure 2 in Castiello, U., Scarpa, M. and Bennett, K. (1995) A brain-damaged patient with an unusual perceptuomotor deficit. *Nature*, **374**, 805–808.

Figure 6.12   After Figure 2.1 in Churchland, P.S. (1986) *Neurophilosophy*. Cambridge: Bradford Books, M.I.T. Press.

Figure 6.13   From Figure 10.6 in Sarnat, H.B. and Netsky, M.G. (1974) *Evolution of the Nervous System*. New York: Oxford University Press.

Figure 6.14   From Figure 26 in Ebbesson, S.O.E. (1980) *Comparative Neurology of the Telencephalon*. New York: Plenum Press.

Figure 6.15   After Figures 374 and 369 in Romer, A.S. (1960) *The Vertebrate Body*. London: W.B. Saunders and Co.

Figure 6.16   From Figure 5 in Ventriglia, F. (1994) *Neural Modeling and Neural Networks*. New York: Pergamon Press.

Figure 6.17   From Figures 7.1 and 7.5 in Damasio, A.R. (1994) *Descarte's Error*. New York: Putnam Books.

INDEX